Naturalists' Handbooks 33

Solitary bees

TED BENTON

Pelagic Publishing
www.pelagicpublishing.com

Published by Pelagic Publishing
www.pelagicpublishing.com
PO Box 874, Exeter, EX3 9BR, UK

Solitary bees
Naturalists' Handbooks 33

Series Editor
William D.J. Kirk

ISBN 978-1-78427-088-9 (Pbk)
ISBN 978-1-78427-089-6 (ePub)
ISBN 978-1-78427-090-2 (Mobi)
ISBN 978-1-78427-091-9 (PDF)

First published in 2017.
Reprinted with minor corrections 2019.

Ted Benton asserts his moral right to be identified as the author of this work.

British Library Cataloguing in Publication Data
A catalogue record for this book is available from the British Library

Cover photographs
Top: Red mason bee, *Osmia bicornis* ♀
Lower left: Long-horned bee, *Eucera longicornis* ♂
Lower right: Hairy-footed flower bee, *Anthophora plumipes* ♀

Printed and bound in India by Replika Press Pvt. Ltd.

MIX
Paper from
responsible sources
FSC® C016779

Contents

Editor's preface

In Britain and Ireland there are about ten times more species of solitary bee than bumblebee and honeybee combined, yet the solitary bees tend to be ignored and we know much less about them. However, they are a fascinating, attractive and diverse group that can be found easily in a wide range of habitats, both urban and rural, and they are important as pollinators. They are easy to study and are certainly no less interesting than bumblebees and honeybees. The main reason for the lack of awareness of solitary bees is probably that for over 100 years and until very recently there was no easily available guide to all the species. The publication in 2015 of a *Field Guide to the Bees of Great Britain and Ireland* by Falk and Lewington at last allowed anyone with an interest to make reliable identifications of solitary bees. However, identification of species can be daunting to the beginner. This *Naturalists' Handbook* provides an introduction to the natural history, ecology and conservation of solitary bees, together with an easy-to-use key to genera, which can act as a stepping-stone to the use of a comprehensive key to species.

I hope that this book will encourage more people to study solitary bees. The more they are studied, the more we shall know about them and the better we shall be able to conserve them for the future. Surprisingly, ex-industrial brown-field sites can provide some of the best nesting sites for solitary bees, which means that solitary bees can be studied and protected, even in the centre of cities.

This *Naturalists' Handbook* on solitary bees is a very welcome addition to the series, complementing the existing titles on bumblebees (no. 6), solitary wasps (no. 3) and ants (no. 24).

William D.J. Kirk
July 2016

Acknowledgements

I wish to thank Nick Owens, Sally Corbet, William Kirk, David Baldock and Mike Edwards for their very helpful comments on all or part of the manuscript of this book, and Nick Owens, Mike Edwards, Bob Seago and Martin Jenner for stimulating my interest, parting with some of their expertise, and sharing some delightful days 'in the field'. Field-work with the late Roy Cornhill was a special pleasure, now much missed. I have benefitted greatly from personal communications with the above, as well as with Peter Harvey, Adrian Knowles, Rob Parker, Graham Stone and David Scott. I am especially grateful to Graham Collins for allowing me to include his excellent keys to the genera of bees of the British Isles. Draft keys to species circulated by George Else have been indispensable, as have several publications, but most especially David Baldock's superb *Bees of Surrey.* For welcoming me to important reserves and generally supporting my activities there, I am particularly grateful to Howard Vaughan and Neil Phillips. Of the organisations that have been vital for this and other works, Colchester Natural History Society, The Bees, Wasps and Ants Recording Society and the Essex Field Club top the list. The editorial contribution of William Kirk and Nigel Massen at Pelagic Publishing has, of course, been exemplary.

Most of all I wish to acknowledge the help of my life companion, Shelley, in setting limits to my obsession with bees, tolerating my 'canine' diversions from the straight and narrow and inordinate delays on our walks, and for never admitting that she manages our small town-centre garden as a bee-paradise.

Keys to the genera of bees of the British Isles

Graham A. Collins thanks the following for their comments on the draft keys: Mike Edwards, Roger Hawkins and Arthur Ewing.

About the author

Ted Benton is emeritus professor of sociology at University of Essex, where he has pioneered the integration of ecological understanding with social theory. He has been an active field naturalist since childhood, and is author or co-author of eight books on entomological topics, in addition to his academic publications and a recent book on Alfred Russel Wallace. His two books in the *New Naturalist* series (*Bumblebees* (2006) and *Grasshoppers and Crickets* (2012)) have both been highly praised. He is hon. President of Colchester Natural History Society, a founder member of the Red-Green Study Group and is involved in environmental campaigning.

About the Naturalists' Handbooks

Naturalists' Handbooks encourage and enable those interested in natural history to undertake field study, make accurate identifications and make original contributions to research.

A typical reader may be studying natural history at sixth-form or undergraduate level, carrying out species/habitat surveys as an ecological consultant, undertaking academic research or just developing a deeper understanding of natural history.

1 Introduction

Everyone now knows at least two things about bees. First, that they provide 'free' services in pollinating a great many of our food crops. Second, that many of them are in steep decline. Thanks to the work of both scientists and conservation movements, these important messages have given bees a high profile in the public imagination. This book is an attempt to take the argument a bit further. The importance of bees as pollinators of a great range of food crops is undeniable. According to some estimates as much as one third of the food we eat is dependent on insect pollination – and bees are by a long way the most effective pollinators. Bee pollination is also important for many species of wild flowers, so, for those of us who love to walk through our few remaining flowery meadows, the bees give us spiritual as well as physical nourishment. But, as I hope to illustrate in this book, bees are also important in their own right: a source of endless fascination and admiration for the wonderful diversity and complexity of their myriad modes of life.

The question of declining bee populations is a much more complicated one. Much of the media coverage equates 'bees' with honeybees. In Britain this means just one species – the domesticated honeybee, *Apis mellifera*. It has been well publicised that beekeepers regularly suffer alarming losses of colonies to disease, most notably as a result of infestations of the *Varroa* mite. Of more international concern has been the phenomenon known as colony collapse disorder. The primary causes of these threats to honeybees are still subject to controversy, but whatever the explanation, the problem points to the risks posed by commercial reliance on just one bee species. Next in line, and very popular with the public, are the bumblebees. Like honeybees, the bumblebees are highly social, and are very effective as pollinators. Because of their unique methods of maintaining their body temperature above the ambient, they can forage earlier, later and in cooler conditions than many other bees. They provide a crucial supplementary pollination service, and are even bred commercially for this purpose. But there has been a great deal of alarm about bumblebee decline, too. The spread of intensive, industrialised agricultural systems, including use of neonicotinoid insecticides, has taken much of the blame for this. The majority of our bumblebee species do, indeed, seem to have declined drastically, although even the worst-affected species may be responding positively to conservation measures, such as agri-environmental schemes. However, a small number of species appear to be coping well with the environmental changes we are imposing on them. These species are often adapted to urban and suburban habitats, to domestic gardens, parks, roadside verges, flood defences and the like. Where these habitats are managed (intentionally or not) to provide a good range of nesting habitats and suitable flowers through from early spring into the autumn, the common bumblebee species seem to thrive.

The honeybee and bumblebees are well publicised and have their place not just in public understanding of their importance as pollinators (or providers of honey), but also in their cultural uses in children's literature, cartoon animations, and as icons of truly selfless social life. However, the rest of this book is devoted to the other bees. That is, the 225 or more species of wild bee that inhabit the British Isles, and which are neither bumblebees nor honeybees. These lesser-known bees are often seen, but only rarely noticed. Nevertheless, some are so familiar that only a brief reminder will call them to mind. One such is the amusingly named hairy-footed flower bee (*Anthophora plumipes*). In common with many of our wild bee species, this bee is more frequent in southern England and Wales, occurring more sparingly into northern England. It is especially common in gardens, and is one of the first bees to appear in the spring. The males emerge first. They are ginger-brown, and look quite like a small bumblebee. However, their behaviour is quite distinctive. They patrol regular routes around flower beds, or flowering shrubs, flying fast, and pausing occasionally to inspect flowers, such as cowslips, rosemary, *Ceanothus* or flowering currant. The point of this behaviour is to locate newly emerged females as they forage for nectar on the flowers, but in most years the males have some days to wait for their first opportunity to mate. Meanwhile, the males, too, need to visit flowers for nectar, and when they do so it is possible to see the reason for their name – a plume of long fine hairs trailing gracefully from their feet. When the females do emerge, they could easily be taken for a quite different species. They are covered in a fine coat of black hair, with a patch of bright yellow-orange hair on each hind leg. This looks superficially like the full pollen-basket of a small bumblebee, and it has a parallel function. The female collects pollen on these hairs (the scopa), and when she is observed at this, we know she has already established her first brood cell. This is constructed from compacted soil or other material, in a hole or crevice in an old wall, or exposed bank. Here she will prime the cell with pollen, mixed with some nectar, before laying a single egg, and sealing it.

Rather later in the year, tidy-minded gardeners might be disappointed to see neatly cut round or ovoid holes in the leaves of their prize roses. This is the work of the appropriately named leaf-cutter bees (*Megachile* species). However, this disappointment is a very small price to pay for the pleasure of sharing one's garden with these fascinating insects. It is a rare treat to watch the female cutting out the shape she needs, and flying off with it to line a brood cell. Soon she will return to the flower-bed to collect pollen among rows of stiff hair on the underside of her abdomen.

Towards the end of June, or early July, the males of another bee take up patrolling routes in the garden. These bees are dark brown, with pairs of yellow marks at the sides of the abdominal segments. The face has a bright yellow pattern, and there are silvery hairs on the feet. There are five small pointed projections at the rear of the abdomen. This is the wool-carder bee (*Anthidium manicatum*). The males patrol in a manner similar to the hairy-footed flower bee, but they are much more aggressive. Each male defends its own patch of flowers, diving at intruding insects – not just other males of its own species, but others, such as honeybees. When in attack mode it bends its abdomen forwards, so its spines become weapons. Again, the point of the patrolling is to locate potential mates, and the males can be seen to hover in front of clusters of flowers, apparently inspecting them for foraging females. When one is located, the male approach dispenses with formalities. He darts at the female, and, if she is slow off the mark, grabs her and mates with her. The females are much more discreet, and it is they who justify the common name of the species. They line their nest cells with 'wool' formed of finely cut plant hairs, which they snip from the leaves of plants such as lamb's ears (*Stachys byzantina*).

This tiny sample of our more familiar solitary bees gives just a hint at the diversity and fascination of the whole group. First, I should give a few words to explain the title of this book. The honeybees and bumblebees are the best known of the bees, and their complex social lives are widely recognised. Most of the bees discussed in this book are termed solitary because they do not have a separate non-reproductive caste of workers. Instead, the females make nests, provide a store of provisions for their offspring and lay their eggs, while the role of the males is generally limited

caste
a form of a species characterised by a particular role in the division of labour, such as queens, workers and drones

to finding and mating with a female. However, for a number of reasons, but often simply because suitable nesting sites are highly localised, large numbers of females – sometimes into the thousands – may nest in close proximity to one another. These dense aggregations can give the appearance of sociality, and the bees certainly don't appear to be solitary! Still, in general, each female recognises her own nest entrance, and there is no overt cooperation between them. To make matters still more complicated, there are several species (some of which are included in this book) that have developed various degrees of social cooperation – possibly as an evolutionary option provided by aggregation in their nest sites. The social lives of some of these species provide interesting evidence bearing on the evolution of sociality. However, there is no simple term to cover the great diversity of bees occurring in the British Isles that are not bumblebees or honeybees. For convenience in this book I will refer to them all as solitary bees, just making it clear where necessary that a few are social.

As the above accounts of familiar species suggest, a local park or garden is a good place to begin to study bees. Some enthusiasts who provide artificial nest-sites as well as the right range of flowers have managed to invite over 100 species into their gardens. But beyond the garden, bees can be found in a great range of habitats – both urban and rural. The open places in woodland (glades, wide rides and edge zones), heaths, old grassland, chalk or limestone grassland, moors, banks, quarries, old walls, roadside verges, railway cuttings, the edges of footpaths, flood defences, sea cliffs and sand dunes all offer nesting and/or foraging resources for bees. Many species make their nests in burrows, and these species are most often found on loose, sandy soils, where the vegetation is sparse. Heather heaths, coastal dunes and worked-out sand and gravel pits are good places to look – especially where there are south-facing bare exposures. These conditions are frequently found in ex-industrial so-called brown-field sites, and some of these are among the richest sites of all, not just for bees but for other invertebrates, too, and for many scarce wild flowers. The value of such sites to wildlife conservation is all too frequently ignored by planners and developers, and many have been lost. Useful work can be done by amateurs in providing the evidence to publicise and defend these much-despised places.

1.1 What is a bee?

Most of us, most of the time, can just intuitively recognise an insect as a bee, a wasp, a hoverfly, or a member of some other group. However, it isn't always so easy to do, as there are some bees that look very much like wasps, and some members of other insect groups that have evolved to look like bees. In some cases there are reasons to think these similarities are not just coincidences. As bees can sting, non-bees that look like bees, but cannot sting, may benefit by being protected from wary predators. Some of the best examples of this mimicry are found among the mimics of bumblebees, but some solitary bees have their mimics, too. So, it might be useful to know in more depth where the bees fit into the general scheme of insect classification, and how to separate them from their various lookalikes.

The bees, together with ants, wasps, horntails, sawflies, velvet ants, parasitic wasps and others, belong to the huge insect order, the Hymenoptera. Almost all species of Hymenoptera have wings, and these are usually transparent or, if tinted, still translucent – unlike, for example, the scale-covered wings of butterflies and moths, or the hard, shell-like modifications of the forewings of beetles. Unlike flies (Diptera - some of which superficially resemble bees), which have just one pair of wings, the hymenopterans have two. However, when an insect is at rest, or, perhaps, taking nectar from a flower, the wings are closed over its back, so it is not easy to tell if it has two pairs or just one. A useful clue in that situation is given by the shape of the head. Bees (and most other hymenopterans) tend to have relatively small compound eyes, situated at the sides of the head and separated by a distinct 'forehead'. In most flies, the head is dominated by very large compound eyes, with little or sometimes no space between them (as seen from above) (Fig. 1.1). Dragonflies and damselflies (Odonata) and mayflies (Ephemeroptera) differ in many ways from the Hymenoptera, but perhaps most obviously their wing-veins form dense networks, contrasting with the much simpler patterns formed by the wing-veins of the hymenopterans (Fig. 1.2).

The Hymenoptera comprises a huge diversity of insects, with numerous subdivisions. The first of these is between the Symphyta and the Apocrita. The Symphyta include the sawflies and horntails, and they differ from the members of the Apocrita in lacking the abrupt narrowing of the body, the 'waist', that characterises the bees, wasps and their allies. But even the Apocrita includes a very diverse range

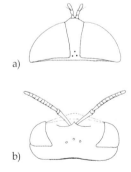

Fig. 1.1 Head of (a) a hoverfly and (b) a bee, viewed from above.

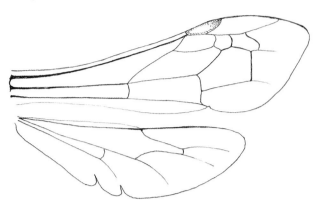

Fig. 1.2 Right forewing and hindwing of a bee, showing simple pattern of veins.

of insects, such as velvet ants, gall wasps, ruby-tailed wasps and parasitic ichneumon wasps, as well as the subgroup to which belong the bees, wasps and ants: the Aculeata.

The aculeates form a coherent grouping of insects, and many entomologists who study bees are also interested in their relatives, the ants and wasps. The Bees, Wasps and Ants Recording Society (BWARS) was set up to foster interest and gather knowledge of these insects in Britain and Ireland, and its website is an indispensable resource for any reader who wishes to take their study of them beyond the limitations of this short book.

But our focus here is on the bees. How do they differ from the ants and wasps? Most obviously, the bees are all furnished with two pairs of wings in the adult stage, whereas most ants are wingless. However, the reproductive castes of ants are winged, and they appear, often in huge numbers, as their nests mature. Like other winged hymenopterans, they have two pairs of wings. These ant castes, and, indeed, all ants, differ from bees in their bodily structure. The differences can be seen quite easily, but some explanation is needed. I mentioned above that the Apocrita (including the ants, bees and wasps) have a marked constriction around the middle of their bodies. This is commonly called the waist. The typical structure of an insect body comprises three main parts – head, thorax and abdomen. In the aculeates the waist *appears* to separate the thorax and abdomen. Unfortunately it is not so simple. In fact, the first segment of what would be the abdomen in other insects is fused with the rear of the thorax to form a structure called the propodeum. To the rear of this, the following segments of the abdomen at first narrow sharply,

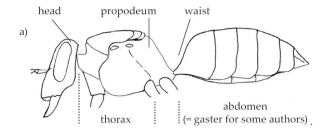

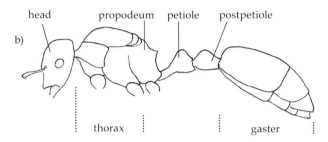

Fig. 1.3 The structure of the body of (a) a bee and (b) a myrmicine ant, viewed from the side.

then widen out again towards the rear of the insect. In the ants, the narrowing of the waist is continued, involving the foremost one or two segments of the abdomen following the propodeum. This gives the appearance of a stalk between the (modified) thorax and the rest of the abdomen. This narrow stalk consists of one greatly narrowed segment (the petiole) in some groups of ants, or two (petiole and postpetiole) on others (the Myrmicinae). The remainder of the abdomen is called the gaster in the case of ants and this term is commonly used to refer to the visible segments of the abdomen in both wasps and bees, too. However, this is not strictly accurate. To avoid unnecessary technicalities the term thorax will be used in this book to refer to the fused middle part of the body, and the term abdomen to refer to the abdominal segments to the rear of the waist (i.e the equivalent of what some authors refer to as the gaster). The term gaster will be used when necessary to avoid confusion and is also used in the authoritative guide to the genera of bees of the British Isles presented in Chapter 8 (Fig. 1.3).

Bees lack the petiole, and the first segment of the (remainder of the) abdomen widens abruptly, to give a more compact impression of the body. In some species, the density of body-hairs is such that the waist is hidden. Many

genus (plural genera)
a group of closely related species

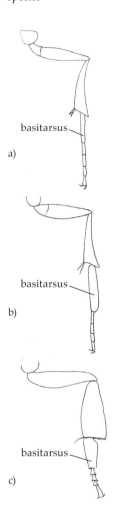

a)

basitarsus

b)

basitarsus

c)

basitarsus

Fig. 1.4 Hind legs of female (a) common wasp, (b) *Nomada* bee and (c) *Andrena* bee showing the shape of the basitarsus.

wasps have a more extended waist, the first one or two segments of the abdomen widening only gradually. These are readily distinguished from bees, but, unfortunately, not all wasps conform. In fact, solitary wasps are believed to be the evolutionary ancestors of bees and some groups, notably the sphecid wasps, are often difficult to distinguish from some groups of bees. One anatomical feature that does seem to distinguish the bees is the structure of the foot, or tarsus. In the bees the first segment of this (the basitarsus) is flattened and wider than the others. This feature is most marked in the case of bees such as those of the genus *Andrena*, whose females collect pollen on their hind legs. However, it is still clear in the wasp-like species of the genus *Nomada*, whose females do not collect pollen (Fig. 1.4). Other distinguishing features of bees either don't work in every case, or require some knowledge of the mode of life of the insect. In both wasps and bees the sexes are most easily distinguished by the number of segments in the antennae (12 in males, 13 in females), but in most species the females have stings – another useful character! Female bees usually (but not always) have some arrangement of hairs, either on their hind legs, under the abdomen or elsewhere, that are used for carrying pollen back to the nest. In general, bees tend to have relatively hairy, or furry bodies, whether male or female. With microscopic examination you can tell that these include many that are plumose – that is, they have fine branches along the length of the hair. Wasps are usually hairless, or have simple hairs on their bodies. Female wasps do not collect pollen, so do not have the hairy adaptations for collecting it.

Another difference between bees and wasps is related to the above. Wasps are generally carnivorous, and the females catch various insects or other invertebrates to feed their larvae. By contrast, most bees feed their larvae on pollen and nectar, and so observation of foraging behaviour can help to decide whether you have a bee or a wasp. Even this is not decisive, however. Several groups of bees that occur in the British Isles have evolved ways of avoiding the housework. They lay their eggs in the nest cells of other bees, and when their larvae hatch they feed on the nectar-pollen mix carefully provided by the original owner of the nest for its own offspring. These 'cuckoo' bees are referred to as cleptoparasites. The females do not have any pollen-carrying hairs, and some of them can be easily mistaken for small wasps. This is especially true of some species in the genus *Nomada*, which have yellow-and-black patterned abdomens.

cleptoparasite
a species that gains its nutrition at some stage in its life-cycle by feeding on the food-stores of another

1.2 Guide to the rest of this book

Chapter 2: Diversity and recognition has two aims. First, it provides an informal account of the diversity of bees present in the British Isles, with brief accounts of the families and genera that are represented here, and some selected species. Second, it aims to enable the beginner to allocate most of the bees anyone is likely to encounter to the right genus. As some genera have only one or two species in the British Isles, that in itself will enable identification of some specimens to species level. In addition, the species selected as representatives of each of the more species-rich genera are chosen in part because they are distinctive enough to be fairly reliably identified, especially when the descriptions are cross-referenced to the photographs. However, it has to be recognised that there will be many specimens that cannot be identified to species level without microscopic examination and extended keys that go beyond the scope of this book.

Chapter 3: Bee lives samples some of the on-going research that is producing a wealth of fascinating information about the life-cycles, nesting behaviour and sexual lives of bees.

Chapter 4: Cuckoos in the nest focuses on the solitary bees that have adopted cleptoparasitic modes of life, laying their eggs in the brood cells of other species.

Chapter 5: Bees and flowers introduces the most-studied aspect of the activity of bees: the mutual adaptations of bees and the flowers they pollinate. The chapter introduces a number of ideas that have been used to map the association of different groups of bees and the flowers they visit, concluding with a brief outline of newer approaches that interpret these relationships ecologically in terms of concepts such as pollination webs.

Chapter 6: The conservation of solitary bees discusses the evidence for conservationists' concern about bee decline, and the alleged pollination crisis. Concern about the potential loss of 'pollination services' has motivated a great deal of research on bees in the agricultural landscape and what measures might maintain their diversity and associated contribution to agricultural and horticultural production. But if our concern is with maintaining bee diversity independently of the services they provide, then the scope must be extended to a wider range of landscapes and habitats, such as urban and suburban green spaces, brown-field sites, downland, heaths and moors and coastal dunes, which act as refuges for bees and other wildlife.

Chapter 7: Approaches to practical work. At several points in this handbook, there have been references to the limitations in our knowledge of the lives of bees and their ecology. This chapter gives a few pointers towards exploratory work that could be undertaken by amateurs.

Chapter 8: Keys to the genera of bees of the British Isles. Chapter 2 provided an informal guide to the different bee genera. Chapter 8 provides the authoritative keys to genera developed by Graham Collins, and reproduced here in its latest version with his permission.

2 Diversity and recognition

This chapter has two main goals. The first is to give the reader an insight into the diversity of life-styles represented among the wild bees that occur in the British Isles, and the way they are classified. The second aim is to provide an aid for the beginner to recognise the different groups, and to be able to identify with reasonable confidence down to species level at least some of the more distinctive species.

Solitary bees have enormous potential for a lifetime of fascination for their students. I hope that as you read further in this book you will be convinced of that. However, there is one big obstacle, and another less serious one. The big obstacle is that there are very many similar-looking species, and so confident identification of them calls for quite detailed study of the anatomical features of a specimen as viewed through the lenses of a binocular microscope. Despite that, there are also many species, often common ones, which are quite distinctive. Once you can name even a small number of species, you are in a position to carry out observational studies on them, and also gain confidence in learning how to recognise others. The less serious obstacle is that the great majority of species have no English names in common use. Some authors have made valuable attempts to reach out to readers who find scientific names challenging by inventing appropriate English names. Where English names are in common use, I have used them here, but in the main I have followed the internationally established practice of using the two-part scientific names. It is necessary to gain confidence in using these if readers wish to take their studies further, access academic research publications and so on. One way to think of these names is by comparison with your own name. The first word in the two-word name is like a surname. It tells you what group of closely related, usually similar, species this species belongs to (its genus). This book should enable you to decide on the first name of all the specimens you find. The second word is like your forename – but in this case tells you which species the specimen belongs to.

A start can be made with common garden species such as the red mason bee, *Osmia bicornis* (formerly *Osmia rufa),* the wool-carder bee, *Anthidium manicatum,* and the hairy-footed flower bee, *Anthophora plumipes.* These are quite distinctive and unlikely to be confused with other species in most areas. However, to take your identification skills further it is necessary to be able to allocate a bee to one of a number of genera. As we shall see, some of these genera include many species that occur in the British Isles, while others have only one or two. In the latter sort of case, if you can tell which genus a bee belongs to you also know the species – or are very close to it. However, some genera, such as *Andrena* and *Nomada*, have many species, which are often very similar in appearance. Even these difficult genera include some distinctive species which can be identified from a good photograph, or by examination with a hand lens (see below). This is especially so if the habitat, season and locality are known, and there is information about behaviour. If the bee is seen entering or leaving a nest, a note should be made of whether this is a burrow in sand or soil, a crevice in a wall, a hole in a fence or a tube in a bee hotel, and any other information. For bees observed visiting flowers, it is often a useful clue to the identity of the bee to make a note of the species of flower. Finally, it is useful to learn the sorts of characteristic features that are used in keys to identify bees, and to make an attempt to show them when relying on photographs.

2.1 What genus does this bee belong to?

This chapter provides an informal guide to the families and genera of bees, together with brief descriptions of some of the more distinctive species in each genus. The descriptions can be matched to the photographic illustrations of most species discussed, and a more definitive check on your allocation of your specimen to genus will be possible by referring to the excellent keys devised by Graham Collins (Chapter 8). For definitive identification of some species it is necessary to make a close microscopic examination of a dead specimen. For some purposes, such as surveying a site to advise on its management, or to defend it from development, this is justifiable, but here an attempt will be made to get as far as possible with identification without killing the insects. As well as photography and close observation, it may be helpful to catch a specimen in a glass or plastic tube. If this is placed in a refrigerator for a short time the insect will become dormant and relevant features can be studied with a hand lens, and photographs taken of relevant anatomical features. In the field without the use of a refrigerator, a piece of tissue can be pushed into the tube and used to gently press the bee against the side of the tube. This inhibits movement and makes it possible to see the relevant anatomical features. The bee can then be released.

abdomen

for simplicity, the term abdomen will be used to refer to the segments to the rear of the waist. See Chapter 1

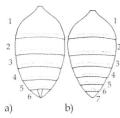

a)   b)

Fig. 2.1 Abdomen of (a) female and (b) male bees, viewed from above.

venation

the arrangement of the veins in the wings

cuticle

the tough outer layer on the surface of the body

*

references cited in the text appear in full under authors' names in References and further reading on p. 186

To allocate a specimen to the right genus it is usually important to find out whether it is a male or female. In the majority of species of solitary bee, the females have areas of pollen-collecting hairs. These may be on the hind legs, under the abdomen, or at the sides of the thorax. However, this is not true of all bees. The female cuckoo bees do not collect pollen, but rely on the work of their hosts. Also there is one genus of nest-making bee (*Hylaeus*) in which the females carry pollen and nectar back to their nests internally. In these groups, the females lack pollen-collecting hairs, and until you are able to recognise them, it is necessary to use other criteria for distinguishing males from females. Males have an extra segment in their antennae (13, against 12 in the females), and also have an extra visible segment of the abdomen – seven in males, against six in females (Fig. 2.1). The females have stings, while the males have complex genital capsules that are usually concealed within the tip of the abdomen (Fig. 7.1).

As we shall see, a wide range of features is used to distinguish the different genera, but the patterns of venation in the forewings are frequently used and it is helpful to record them photographically when you have a captive bee. Fig. 2.2 shows the pattern of forewing veins, the significant ones given names that I will use in this chapter. Figs. 2.3 through to 2.10 give the main variations that can be used as a guide in allocating a bee to its correct genus. The colour and distribution of hair over the insect's body, the colour of the cuticle, any red, white or yellow markings, the shape of the abdomen and presence or absence of spines or other projections on various parts of the body may all help in identification.

For those who are inspired to take their study of these insects further, the website of the Bees, Wasps and Ants Recording Society is an invaluable resource, as is their regular newsletter. We now have two comprehensive and beautifully illustrated guides to the bees of the British Isles (Falk and Lewington 2015 and Else and Edwards 2018*). Baldock's *Bees of Surrey* (2008) includes excellent accounts of a high proportion of the British species and colour photographs of a good selection of those. Also useful are several general photographic guides to insects, such as Chinery (2005) and Brock (2014). A remarkable work on the natural history of a Surrey garden (Early, 2013) includes fine photographic illustrations and information on familiar garden species. Scotland, although it has fewer species than more southerly parts

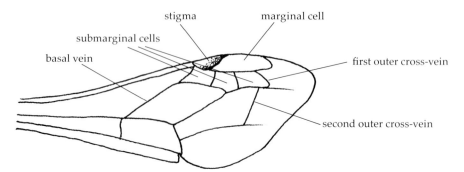

Fig. 2.2 Right forewing of a solitary bee, showing the main features used to distinguish genera.

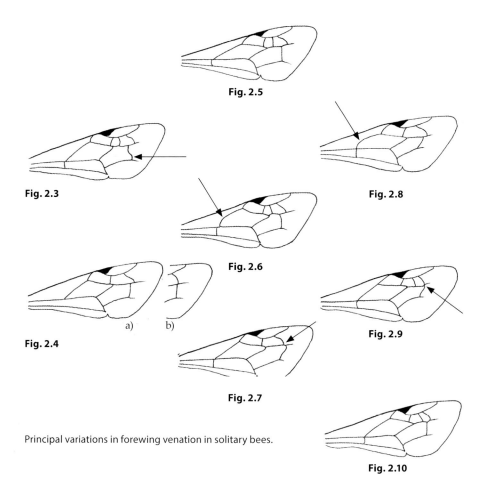

Principal variations in forewing venation in solitary bees.

of the British Isles, does have a distinctive bee fauna, some species being better established in Scotland than elsewhere. Edwards (2013) provides an excellent introduction to the study of Scottish bees. Many local natural history societies or wildlife organisations will have one or more experienced bee-specialists, who may be able to help with a reference collection and guidance in the use of identification keys.

2.2 Bee classification

Some authors include all bees in the family Apidae, with six or seven subfamilies. Others raise these subfamilies to the status of families, including one of the subfamilies in a more narrowly defined family Apidae. This leaves six families, and is the approach adopted here. The six families are:

Colletidae, comprising just two genera represented in the British Isles: *Colletes* and *Hylaeus*.

Andrenidae, again comprising just two genera: *Andrena* and *Panurgus*.

Halictidae, including *Halictus*, *Lasioglossum*, *Dufourea* and the cleptoparasitic *Sphecodes*.

Melittidae, including *Melitta*, *Dasypoda* and *Macropis*.

Megachilidae, including *Megachile*, *Osmia*, *Hoplitis*, *Chelostoma*, *Heriades*, *Anthidium* and the cleptoparasites in the genera *Stelis* and *Coelioxys*.

Apidae, including *Anthophora*, *Ceratina*, *Xylocopa*, *Eucera* and the cleptoparasites in *Nomada*, *Epeolus*, and *Melecta*, as well as the bumblebees (*Bombus*) and the honeybee (*Apis*). The last two genera are, of course, not covered in this book.

Family Colletidae

Genus *Colletes*

The nine species belonging to genus *Colletes* are medium-sized mining bees. The females nest in burrows in the ground, often being associated with steep or vertical exposures with sandy soil, or in banks or in holes in walls. They are sometimes referred to as plasterer bees, as they secrete a cellophane-like material to line their cells. Most species collect pollen from a narrow range of flower species, including heather, ivy or plants in the Asteraceae (daisy family). All species except one have a distinctive pattern of ginger-brown hair on the thorax and dense bands of flattened whitish or yellow hairs at the rear margins of the abdominal segments. Their tongues are short and bi-lobed at the tip, and, viewed from the front, the inner margins of the eyes converge below, giving a roughly triangular

cleptoparasite

a species that gains its nutrition at some stage in its life-cycle by feeding on the food-stores of another

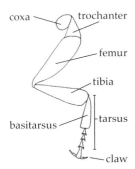

coxa — trochanter

femur

tibia

basitarsus — tarsus

claw

Fig. 2.11 The leg of a bee.

marginal area
rear portion of a dorsal
plate (tergite) of a bee
abdomen, usually visibly
distinct from the rest of
the plate

abdominal tergite
the dorsal (upper)
surface of one of the
abdominal segments

shape to the face. The pattern of veins in the forewing is distinctive: three submarginal cells, and the second outer cross-vein bulging out in its lower half (Fig. 2.3). Confident identification of the species requires microscopic examination. However, for those species that are associated with particular habitats, or collect pollen from a narrow range of flower species, this gives a useful clue to their identity.

Colletes succinctus (Fig. 2.23). This species is a heather specialist, occurring on heaths and moors throughout the British Isles. However, in some localities (e.g. Kerry, Ireland) the females collect pollen from yellow-flowered Asteraceae such as ragwort (Edwards & Telfer, 2001). Habitat is a useful clue to identification, and females have the middle part of the rear margin of the first abdominal tergite orange-tinted and semi-transparent. In females, there are some black hairs among the pale ones on the dorsal edge of the tibia (Fig. 2.11) of the hind leg, but these are not easy to see in photographs.

Colletes halophilus (Fig. 2.24). This is a species of coastal dunes and estuarine habitat, sometimes common along the coast of East Anglia, northwards to Yorkshire, and on parts of the south coast of England. The marginal area of the first abdominal tergite is as in the previous species, but the pollen hairs on the hind leg are all pale. *Colletes hederae* (Fig. 2.25) is a very similar species, but it is on average larger than *C. halophilus*, and flies rather later, as the females collect pollen from ivy.

Colletes cunicularius (Fig. 2.26) was formerly restricted to north-western coastal dunes in England and west Wales, but has since been discovered on some inland sandy localities, including some in East Anglia and the south coast of England. It has the characteristic wing venation and facial features of *Colletes*, but lacks the dense bands of hair on the abdominal tergites. It flies in April and May and the females collect pollen from willows (*Salix* species).

Genus *Hylaeus*
There are 12 species of *Hylaeus* currently known in the British Isles. They are small to medium-sized, mainly black, sparsely haired bees. Unlike the mining bees, they do not nest in burrows but utilise hollow, dead stems of plants such as bramble, or crevices and holes in walls or wood. As in *Colletes*, the nests are lined with a cellophane-like material secreted by abdominal glands. The females have no scopa or pollen basket, but carry nectar and pollen back to their

nests internally. They are sometimes known as yellow-faced bees, as all but one species have pale yellowish markings on the cuticle of the face. These markings are usually more extensive in the males, and there are often yellow areas on the legs. There are just two submarginal cells in the forewing (Fig. 2.4). In some species the second outer cross-vein is bowed outwards. There could be confusion with males of *Macropis*, which also have extensive yellow markings on the face. However, *Macropis* males are larger, more robust-looking insects than *Hylaeus*, and have enlarged hind femora and tibiae (Fig. 2.12).

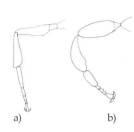

Fig. 2.12 Hind legs of (a) *Hylaeus* and (b) *Macropis* males.

Hylaeus communis (Fig. 2.27). This is one of the commonest species in the genus, and is widespread in England and Wales. There are also some records from Scotland and Ireland. The distinctive yellow facial marking in the male curls round the bases of the antennae. The females are best identified by the *absence* of patches of flattened white hairs at the sides of the first abdominal segments (these are present in other *Hylaeus* females except *pectoralis*). They have roughly triangular yellow markings on the face, much smaller than those of the males.

Hylaeus pectoralis (Fig. 2.28). This species has a very restricted distribution in wetland habitats in south-coastal counties of England, and inland in East Anglia. The females make their nests in galls on common reed. In both sexes the first abdominal tergite is unpitted and shiny.

Hylaeus hyalinatus (Figs. 2.29, 2.30). This is a common and widespread species in the British Isles, often inhabiting gardens. The yellow facial markings are extensive in the males, but reduced to two elongated triangles in the females. The yellow areas on the legs are extensive, especially in the males, and the first abdominal tergite is pitted but shiny between the pits.

Family Andrenidae

Genus *Andrena*

This is a very large and diverse genus (almost 70 species) of mining bees. Identifying individuals down to species level is difficult even for the experienced bee-watcher. They vary from as little as 6 mm in length (male *Andrena minutula*) to as much as 17 mm (female *Andrena hattorfiana*). The females make their nests in burrows which they dig into bare or sparsely vegetated flat or sloping ground, often on footpaths, and sometimes in very large aggregations. The brood cells are lined with a waxy material. Several species are among the first to be seen in early spring, and this is a good time of year to begin learning how to identify them. Some species have later flight periods, and some of the spring species have a second, summer emergence. The cuticle is black, but in some species there are reddish patches, especially on the abdomen, and in a few species the males have cream or whitish markings on the face. The forewing has three submarginal cells, the outer one longer than that in the middle. The basal vein is straight, or gently curved, and the second outer cross-vein is more-or-less straight (i.e. it does not bulge out as in *Colletes*) (Fig. 2.5). The inner margins of the eyes are parallel, and there are narrow troughs (foveae) with fine, velvety hair on the face and running close to the inner margins of the eyes. These are often just visible on sharply focused photographs. The tongue is short and pointed at the tip. The females have pollen-collecting hairs on the hind legs (and sometimes at the sides of the mid-body), not under the abdomen. Confusion is possible between *Melitta* and *Andrena*, as both genera have similar patterns of venation in the forewings. *Melitta* lacks facial foveae and has obliquely cut-off tips to the antennae. However, the most easily recognised character is the relatively much wider final segment of the feet (tarsi) in *Melitta*, compared with *Andrena* (Fig. 2.13). However, these features are not always easy to discern on live specimens, and so it is useful to learn to identify the *Melitta* species independently. In general, females of *Andrena* are more distinctive than males, so it is advisable to begin with females when embarking on the difficult struggle to learn the species of *Andrena*.

The tawny mining bee, *Andrena fulva* (2.31). This is perhaps the most familiar and distinctive species in the genus. The female is densely clothed with bright reddish hairs on the dorsal surface (upperside), giving way to black hairs on the ventral surface (underside). The legs and face are also black with black hairs. The males are smaller and less brightly coloured, with ginger hairs on both dorsal and ventral surfaces. The covering of hairs in the males is also less dense than in the females, and the hairs on the face are pale, with a few black hairs close to the inner margins of the eyes. The females make their nests in burrows in short grassland, including lawns, and the species is widespread in England and Wales, becoming scarcer northwards into Scotland. The flight period is from late March to June, with only a single annual brood.

Andrena cineraria (Fig. 2.32) This is another very distinctive and widespread species. In the female the

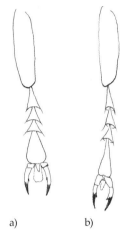

a)   b)

Fig. 2.13 Tarsal segments of (a) *Melitta* and (b) *Andrena*.

cuticle is black with bluish tints, and there are bands of grey-white hairs at the front and rear of the thorax, and similarly coloured hairs on the face. The male is smaller and with more extensive pale hairs. The females make their nests in burrows, often in short grassland such as lawns and parks, often in the same habitats as the previous species. It has a scattered distribution across England and Wales, and in Ireland. There are older records from Scotland. It is on the wing from late March until June.

Andrena haemorrhoa (Fig. 2.33). Another widespread mining bee of early spring, this species can often be seen foraging from the catkins of sallow (*Salix* species) and, later, on the blossom of fruit trees. It is one of several species in which the females have ginger-brown hairs on the thorax, and mainly black abdomen. However, the combination of white hairs on the face and ventral surface, orange cuticle of the hind tibiae and foot-segments, and the fringe of ginger-red hairs at the tip of the abdomen will help to distinguish it. The males are smaller and less densely hairy, but have a similar colour pattern, except that the white on the face and underside is replaced with ginger-orange. They have a distinctive dark blotch on the outer surface of the otherwise orange hind tibia.

Andrena clarkella (Fig. 2.34). This is another early spring species, often found in open woodland. It too has the pattern of ginger-red thorax, and black hairs on the abdomen. The hind tibiae, foot-segments and pollen hairs in the female are orange. However, the hairs on the face and ventral surface are black, and there are no ginger-red hairs at the tip of the abdomen. The females dig their burrows in banks, on flat bare ground, or in root-plates of fallen trees. This species is local, but widely distributed across the British Isles, including Scotland and Ireland.

Andrena florea (Fig. 2.35). This species flies later, and is most likely to be seen in June. It has a very restricted distribution in the south-east of England, and here collects pollen from white bryony (*Bryonia dioica*) only. Where they occur they can best be located by watching for them by patches of this plant. However, the bryony flowers are popular with other species of insects too, so care is needed. Both sexes have narrow bands of red across tergites 2 and 3 of the abdomen.

Andrena hattorfiana (Fig. 2.36). This large and distinctive species is quite widespread but very localised in southern England, south Wales and the East Anglian Brecks. The females collect pollen from field scabious, small scabious and greater knapweed, and can often be identified by a combination of their large size and the pink scabious pollen on the hairs of the hind legs. However, the very similar, but smaller *Andrena marginata* (Fig. 2.37) is also closely associated with scabious, and the two species sometimes occur together.

Genus Panurgus

In the British Isles there are only two small and rather inconspicuous species. Both are mining bees, which line their nests with waxy material. Both species favour dry, sandy grasslands and heaths, and are associated with yellow-flowered Asteraceae (daisy family). The forewing has two submarginal cells, the basal vein is straight or gently curved, and the second outer cross-vein joins the median vein closer to the wing-base than does the first outer cross-vein (Fig. 2.4). The face is black, with black hairs, and the tongue is short and pointed at the tip. The legs are black and there is an arolium between the claws (Fig. 2.16). The females have pollen-collecting hairs on the hind legs.

arolium (plural arolia)
a small pad or projection between the claws on the final segment of the foot

Panurgus banksianus. This medium-sized bee is thinly covered in black hair, with yellowish pollen hairs on the hind legs of the female. The male is similar, but with golden hairs on the hind and midtibiae and tarsi. They fly from June to August, and are found on grasslands on sandy soils, where the females collect pollen from yellow-flowered Asteraceae. The males are noted for curling up in the flowerheads late in the day or in dull weather. The species is well distributed in south-eastern counties of England, but is mainly coastal further west, in Wales and East Anglia (but also inland in the Brecks).

Family Halictidae

Although there are four genera represented in the British Isles, one, *Dufourea*, has only two species that have been recorded here, both of which are believed to be extinct. Of the others, two, *Lasioglossum* and *Halictus*, are generally small species, with mainly black or metallic-coloured cuticle, and with pollen-collecting hairs on the hind legs in the females. All are mining bees, and several are social. Species of the fourth genus, *Sphecodes*, are cleptoparasites, usually invading the nests of other bees in the same family, but also, less commonly, some *Andrena* species. Most *Sphecodes* species have extensive areas of red on the abdomen, and the females lack pollen-collecting hairs. All three extant genera share a common pattern of forewing venation. This is similar to *Andrena* and *Melitta*, with three submarginal cells and an approximately straight second outer cross-vein. However, the basal vein is strongly arched as it approaches the longitudinal vein, meeting it almost at a right-angle (Fig. 2.6).

Genus *Halictus*

There are now only six species in the British Isles (including two that occur only in the Channel Islands), and of these just two are relatively common and widespread. Both may be solitary or social, depending in part on climate and latitude. The cuticle is mainly black or with metallic reflections and there are flattened white hair bands on the rear margins of some of the abdominal tergites.

Halictus rubicundus (Fig. 2.38). This is a relatively large species (up to 10 mm), with a black cuticle and bright yellow-orange tibiae and tarsi in the female, and completely yellow-orange legs in the male. There are dense white hair-bands, some of them broken in the middle, on the rear marginal areas of the abdominal tergites. It is common and widespread throughout the British Isles, including Scotland and Ireland.

Halictus tumulorum (Fig. 2.39). This species has distinctive metallic reflections from the cuticle, especially in the females. These have dense white hair bands on the marginal areas, but also narrow bands on the basal (front) areas of tergites 2 and 3 of the abdomen. The females are active from March onwards, with males appearing at the end of June or early July. They are common and widespread in southern Britain becoming more sparsely distributed northwards to Scotland.

Genus *Lasioglossum*

The 33 species currently known from the British Isles are often very difficult to identify, even with microscopic examination. All species are ground-nesting, often in large aggregations, and some are social, with distinct worker and queen castes. The males often have some yellow markings on the face or legs, and some red on the abdomen. They are very similar to *Halictus* species though many (but not all) species have bands or lateral patches of white flattened hair at the front of some abdominal tergites (the rear in *Halictus*). Although the pattern of wing venation is similar in the two genera, the outer cross-veins are finer and slightly less well-marked than the rest of the veins in *Lasioglossum* (undifferentiated in *Halictus*). This feature is not easy to see, and is clearer in females.

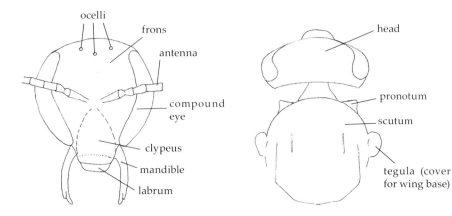

Fig. 2.14 The face of *Lasioglossum calceatum* (male).

Fig. 2.15 Head, pronotum and scutum of *Lasioglossum malachurum*, viewed from above.

Lasioglossum calceatum (Fig. 2.40) The cuticle is mainly black, but often with reddish areas on the abdomen in males. In females the rear margins of tergites 1 to 4 are translucent, and there are wedge-shaped patches of flattened white hairs laterally on the second or second and third abdominal tergites. Body-hairs, including the pollen-hairs on the hind legs, are yellow. The males have a rather enlongated face, with white hairs, and yellow cuticle towards the tip of the clypeus (Fig. 2.14). The abdomen is elongated, with anterior patches of flattened white hairs on tergites 2 to 4, sometimes forming a complete band. The tarsi are yellow, shading to pale brown. It is on the wing from April to October, and is both common and widespread in the British Isles, including Scotland and Ireland. In summer and autumn it is commonly seen in parks and gardens on a range of flowers, but especially Asteraceae. Its nests are parasitized by *Sphecodes monilicornis*.

Lasioglossum malachurum (Fig. 2.41). The cuticle is dark brown to black, and in both sexes the dorso-lateral corners of the pronotum form a right-angle, giving the appearance of square shoulders to the front of the thorax (view from above) (Fig. 2.15). The females have golden-yellow pollen-hairs on the hind legs, and bands of golden hair at the rear of the abdominal segments, as well as small paired wedge-shaped patches of flattened white hair towards the front of tergites 2 and 3. The males have small, inconspicuous wedges of white hair on tergites 2 and 3, and yellow legs with black patches on the tibiae. This species is common in south-eastern England and is extending its range. It forms

pronotum
the foremost section of the thorax

aggregations on bare ground, often at the sides of footpaths, and is social. Fertile females over-winter and establish nests in the spring, producing a first brood of workers, followed by a summer brood of males and fresh females (gynes).

Lasioglossum morio (Fig. 2.42). This is a very small bee (not more than 6 mm in length), with dull metallic blue-green reflections from the cuticle. It is very common and widespread in southern England, but becomes more scattered northwards, reaching as far north as Cumbria and North Yorkshire. Like the previous species, it is primitively social, with females active from early spring, but males not making an appearance until late June or July. They continue on the wing until early autumn.

Genus *Sphecodes*

17 species of this genus of cleptoparasitic bees are present in the British Isles (including one that occurs only in the Channel Islands). The cuticle is mainly black, and very sparsely covered with short silvery hairs. The dorsal surface of the thorax (and often the head) is shiny with distinct punctures. Females can be distinguished from those of other genera by the combination of extensive orange-red areas on the abdomen and lack of pollen hairs on the hind legs. The forewing venation is similar to that of *Lasioglossum* and *Halictus* (Fig. 2.6). Males often have less orange-red on the abdomen, but can be distinguished by the combination of the wing venation together with convex ventral surfaces to the antennal segments, giving the antennae a 'knobbly' appearance. Unfortunately, the species are very similar to each other in general appearance, and in most cases can only be confidently identified by microscopic characters.

Sphecodes ephippius (Fig. 2.43). This is a common and widespread species which is on the wing from April onwards through the summer. The females lay their eggs in the nests of several common species of *Lasioglossum*. The first three abdominal tergites are bright orange-red, the remainder black.

Sphecodes monilicornis (Fig. 2.44). This species also has a predominantly orange-red abdomen, but in the female is distinguished by the shape of the head: viewed from above, the sides of the head are approximately parallel before they turn inwards. The hosts include other members of the Halictidae, notably *Lasioglossum malachurum*, *Lasioglossum calceatum*, and *Halictus rubicundus*. The species is widespread throughout the British Isles.

Sphecodes pellucidus. This is another common and

widespread species in England and Wales, reaching as far north as southern Scotland, and with records from the east coast of Ireland. It takes *Andrena barbilabris* as its host, and may be seen at nesting aggregations of that species.

Sphecodes puncticeps (Fig. 2.45). This species is also common and widespread in southern England and Wales, reaching as far north as Cumbria in the west. The females lay their eggs in the nests of several *Lasioglossum* species.

Family Melittidae

This is a small but diverse family of mining bees. There are three genera to be found in the British Isles.

Genus *Macropis*

There is only one British species, *Macropis europaea* (Fig. 2.46). The flight period, from mid-July to early September, is synchronised with the flowering of yellow loosestrife, from which the females forage for pollen and floral oils. The females nest in burrows and use plant oils for the lining of their brood cells. The species is local and confined to southern counties of England and parts of East Anglia. Both males and females are closely associated with yellow loosestrife, which helps in identification. They are small and black, superficially similar to *Halictus* or *Lasioglossum*, but there are just two (not three) submarginal cells in the forewings (Fig. 2.4). There are dense white hair-bands on the rear margins of some gastral tergites in both sexes, but these are absent from the first two tergites. Females have very distinctive adaptations for carrying pollen and oils on their hind legs: the pollen-collecting hairs on the hind tibia are white, and those of the hind basitarsus (first segment of the foot) are black. Even when a female has a full load, her habit of holding both back legs out to the side is quite distinctive (Fig. 2.46). The males have yellow faces, but can be distinguished from other pale-faced males, such as those of *Hylaeus* species, by their larger size and by the strongly expanded hind femora and tibiae (Fig.2.12).

Genus *Dasypoda*

There is only one species in the British Isles, the hairy-legged mining bee, *Dasypoda hirtipes* (formerly *Dasypoda altercator*) (Fig. 2.47). The females burrow into sandy soils, especially on heaths or coastal dunes, and are very locally distributed in southern England, East Anglia and coastal Wales. They visit a range of Asteraceae, such as ragwort, hawkbit and sow-thistle and are on the wing from the end of June until early September. The females are quite distinctive, large (15 mm), hairy, with alternating bands of golden-brown and black hairs on the abdominal segments, and exceptionally long drapes of pollen-hairs on the hind tibiae and tarsi. They visit flowers in the mornings, and are active in their burrows later in the day. The males also have long hairs on the body and legs, but these are less brightly coloured than those of the female, varying from pale brown to grey-white. The wing-venation is as shown in Fig. 2.4 (two submarginal cells, the second outer cross-vein joining the median vein closer to the wing-base than does the first outer cross-vein).

Genus *Melitta*

There are four species in the British Isles, but one of these (*Melitta dimidiata*) is extremely localised. As all four species collect pollen from just one or a narrow range of flower species, this is the best way to locate them. The forewing venation is very similar to that of *Andrena* (Fig. 2.5), but there are two features that readily separate them from members of that genus: the final segment of the foot is relatively wide (narrow and elongated in *Andrena*) (Fig.2.13) and the antennae have a flat, oblique tip. Unfortunately the last-mentioned character can only be seen from some angles, so may not show up in a photograph. There are distinctive facial features, too, but these are often not visible in photographs. The females have pollen-hairs on the hind tibiae and basitarsi, and both males and females have white hair-bands on the marginal areas (rear sections) of the abdominal tergites.

Melitta haemorrhoidalis (Fig. 2.48). The females nest in burrows and collect pollen from *Campanula* flowers. They are found on calcareous grassland but also in gardens where they visit cultivated campanulas. They fly from mid-July to late August and have a restricted distribution in the British Isles, occurring mainly in southern and south-eastern England, with a few outlying records in Wales, East Anglia and further north. The females have distinctive ginger-red hairs at the tip of the abdomen.

Melitta leporina (Fig. 2.49). This is another ground-nesting bee. The females collect pollen mainly from vetches and clovers (family Fabaceae). Their flight period is July and August, and their UK distribution is similar to that of the previous species, though they do not extend so far to the north. Both sexes have pale yellow or whitish hair bands on the rear margins of the abdominal tergites.

Melitta tricincta (Fig. 2.50). This species is very similar to the previous one, and is best detected by its association with red bartsia (*Odontites verna*), from which the females collect pollen. It can be observed darting rapidly from one plant to another where red bartsia is common. It flies from late July to August, when red bartsia is in flower, but has a restricted southern and south-eastern distribution in England, extending into East Anglia.

Family Megachilidae

This family is represented by eight genera in the British Isles, two of them comprising cleptoparasitic species. As is usual with cleptoparasites, the females of these species have no pollen-collecting apparatus, but the females of all the others are distinctive in collecting pollen among long, semi-erect brushes of hairs on the underside of the abdomen (abdominal sternites). Most species nest in hollow plant stems or other cavities such as old beetle holes, but a few nest in burrows in sandy soils. All genera have just two submarginal cells in each forewing.

Genus *Anthidium*

There is only one species in the British Isles. The wool-carder bee, *Anthidium manicatum* (Fig. 2.51). This is one of the few species in which the males are usually larger than the females. The larger males aggressively defend territories, and females line their nests with clipped plant hairs. Both sexes have dark brown cuticle, with distinctive paired yellow spots on the first five abdominal tergites, and yellow markings on the face (more extensive in the males). The males also have distinctive long white hairs on the tarsi, and five sharp spines projecting from the rear two segments of the abdomen. They are most likely to be seen in parks and gardens as well as marginal land, woodland and coastal sites. They are on the wing through June and July, and are widely distributed through south and central England, being mainly coastal in Wales and in western Scotland. They appear to be extending their range and have recently arrived in Edinburgh.

Genus *Megachile*

There are currently just seven species in the British Isles. They are medium to large-sized bees, which line their brood-cells with cut segments of leaves or petals – hence their common name leaf-cutter bees. The abdomen appears parallel-sided, giving the bee a rather stocky, oblong shape. The mandibles, especially in the females, are wedge-shaped

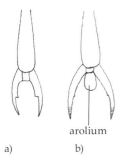

arolium

a)   b)

Fig. 2.16 Final tarsal segments of (a) *Megachile* and (b) *Osmia* showing absence or presence of an arolium between the claws.

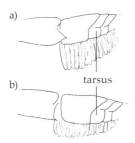

a)

b)   tarsus

Fig. 2.17 Fore tibia and tarsal segments of (a) *Megachile willughbiella* and (b) *Megachile maritima*, males.

with several teeth along the cutting edge. The tongue is relatively long, and there is no arolium between the claws. *Osmia* is very similar but *Osmia* species do have an arolium between the claws (Fig. 2.16). There are two submarginal cells in the forewing, and the second outer cross-vein meets the median vein slightly closer to the base of the wing than does the first outer cross-vein (Fig. 2.4b). Distinguishing species can be difficult, but there is an accessible published key (Else, 1999).

Megachile willughbiella (Fig. 2.52). This is one of the largest species in the genus (females approximately 15 mm in length), and is widespread in urban areas, including gardens, reaching as far north as northern Scotland, and with records from the south-west of Ireland. The flight period is from mid-June to early August. The males have distinctive wide, pale front tarsi that look as though the insect has boxing gloves on. These are used in mating (see Chapter 3). The similar, but usually southern and coastal, *Megachile maritima*, also has flattened front tarsi, but the shape is slightly different and the former species has much longer hairs on the ventral (lower) edge of the tarsi (Fig. 2.17). The females have very narrow bands of white hairs along the rear edges of their abdominal tergites. The pollen-hairs of females are golden except for those on the final two sternites, which are black (except in Ireland, where the black hairs are fewer or even absent).

Megachile centuncularis (Fig. 2.53). This is another familiar leaf-cutter of gardens, and is also widespread in the British Isles, reaching as far north as the Scottish Highlands. It also occurs in the east of Ireland. The males do not have the flattened fore tarsi of the previous species, and the females have golden pollen-hairs throughout, except for a narrow fringe of short black hairs at the rear.

Megachile leachella (formerly *Megachile dorsalis*) (Fig. 2.54). This is a smaller species which is found on coastal sand dunes and some sandy inland sites in southern Britain as far north as the Wash in the east, and north Wales in the west. The females have pure white pollen-hairs, and two distinctive white spots on the dorsal surface of the final abdominal segment. This segment is densely clothed by flattened white hairs in the male. Both sexes have green eyes.

Genus *Osmia*

There are 12 British species, several of them familiar in gardens. Three specialise in nesting in snail shells, while the rest are quite catholic in their selection of nest-sites, which include plant stems, cavities in old walls and burrows in the ground. The cell divisions and plugs at the entrance to their nests are made from mud or chewed plant fragments. They are similar in general appearance to *Megachile* species, but unlike them have arolia between their claws. The pattern of forewing venation is very similar to that in *Megachile* (Fig. 2.4).

The red mason bee, *Osmia bicornis* (formerly *Osmia rufa*) (Fig. 2.55). This is one of the few species with a commonly used vernacular name. It is very frequent in gardens, and readily makes its nests in artificially provided lengths of cane or glass tubes. The females have a pair of distinctive forward-pointing projections from the face and a golden-red pollen brush on the underside of the abdomen. Males are much smaller with a fine covering of long golden hair on the abdomen. It is common in south and south-central England and south Wales, but more patchily distributed further north to Scotland.

Osmia caerulescens (Fig. 2.56). This species is double-brooded, flying from April or May through June, and again in July and August. It is fairly common in southern and south-eastern England, and with a scattered distribution further north. It frequents open woodland and gardens and nests in cavities such as plant stems and beetle holes in dead wood. There are strikingly beautiful metallic blue-purple reflections from the cuticle in females, and bronze-green in males. The pollen-collecting hairs on the underside of the female's abdomen are black, and the rather scanty body-hair is white or very pale yellowish.

Osmia leaiana (Fig. 2.57). This is another fairly frequent inhabitant of parks and gardens, and has a similar distribution to the previous species. The females have blue metallic reflections from the cuticle and ginger-brown hairs on the face, dorsal surface of the thorax, and in bands on the abdomen. The pollen-collecting hairs on the underside of the abdomen are orange. Males are very similar to males of the previous species, but the front (declivity) of the first abdominal segment (Fig. 2.1b) is matt (shiny in *O. coerulescens*). Unfortunately, this feature cannot be seen in photographs of living specimens.

Osmia bicolor (Fig. 2.58). This species is rather locally distributed, south of a line from the Bristol Channel to the Wash, and along the coast of south Wales. The females are black with black hair on the head and thorax, but bright ginger-red hairs on the abdomen, including the pollen-hairs, and lower leg segments. The males are smaller, and black with ginger hairs. This is one of three *Osmia* species that nest in snail shells, and females build a tent-like structure over their shell-nests.

Osmia spinulosa (Fig. 2.59). This small species was formerly included in the genus *Hoplitis,* and is very similar to the following species, except that the females have yellow (not white) pollen-brushes on the underside of their abdomen. The females collect pollen from flowers in the Asteraceae, and make their nests in snail shells. Males have a sharp spine projecting from the underside of the first abdominal segment (sternite) which is distinctive but unfortunately not visible in photographs. The bee is often common in rough grassland or open forest rides, but its distribution is restricted to southern and south-eastern England, East Anglia and south Wales, with a few outlying records from further north.

Genus *Hoplitis*

There is currently only one British species, *Hoplitis cla-viventris*, which resembles *Osmia spinulosa*, but the pollen-hairs on the ventral surface of the female abdomen are white, not yellow as in that species, and the males lack the ventral spine on sternite one (though they do have a blunt projection on sternite two). This bee is generally uncommon but quite widely distributed in southern Britain, reaching as far north as Cumbria, and is on the wing from late May to the end of August.

Genus *Heriades*

Only one species* is established in the British Isles, *Heriades truncorum* (Fig. 2.60). This is a small black bee, sharing many characteristics with *Osmia* and *Hoplitis*. The females have pale yellow-orange pollen-hairs, and narrow pure white hair fringes on the rear margins of tergites 1 to 5. The cuticle is black and shining, with dense punctures. The male is similar, but without the pollen-hairs. There is a distinctive ridge at the front of the first abdominal tergite, but this is obscured by the wings when the insect is settled and so does not show up in photographs. The forewing venation is similar to most other genera in this family (Fig. 2.4). When females are foraging they vibrate their abdomens in a very distinctive way. They nest in beetle holes in dead wood and sometimes in other small cavities. They collect pollen from Asteraceae, and are

* Postscript (2019)
Another species, *Heriades adunca,* has recently been discovered by the Thames in east London

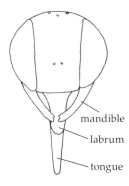

Fig. 2.18 Face of *Chelostoma*.

Fig. 2.19 Final segments of the abdomen of male *Chelostoma*, viewed from behind.

megachilid
belonging to the family Megachilidae

most frequently reported as foraging from ragwort. This has been considered a scarce species of south-eastern England, but seems to be rapidly expanding its range.

Genus *Chelostoma*

There are just two species that occur in the British Isles. Both are small to medium sized, slender, black, and with white pollen-hairs. The genus is distinguished by the facial features: relatively long mandibles that meet below the clypeus when the insect is at rest, revealing an elongated labrum (Fig. 2.18). Males have a two-pronged projection from the rear of the abdomen (Fig. 2.19). *Chelostoma florisomne* has a scattered distribution in England and Wales, reaching northwards almost to the Scottish borders, but is more common in south-eastern England. Pollen is collected from buttercups, a useful method of locating the bee. *Chelostoma campanularum* (Fig. 2.61) is a very small species which has a more restricted distribution, also centred on the south-east of England. The females collect pollen from campanulas, including garden varieties.

Genus *Coelioxys*

There are seven species in the British Isles, all of which are nest-parasites of species of *Megachile* or *Anthophora*. Even where the hosts are common, *Coelioxys* species are relatively rarely seen. In all but one species, the female is very distinctive in shape. It has a conical abdomen which narrows to a rigid point at the rear. This is used to open the host's brood cells. The male's abdomen is more blunt-tipped and bears several backward-pointing spines. The black shiny cuticle is marked by coarse punctures, especially on the head and thorax, and there are narrow bands or patches of dense white flattened hairs on both surfaces of the abdomen. Microscopic examination is needed to identify them confidently to species but one species, *Coelioxys elongata* (Fig. 2.62), is frequently seen in gardens and other habitats where one of its main hosts, *Megachile willughbiella,* is common. It is on the wing from June through to August, coinciding with the flight period of its hosts. It is widespread in the British Isles, including Ireland, but it becomes predominantly coastal northwards to Scotland.

Genus *Stelis*

These small to medium-sized black bees are cleptoparasites of various other megachilid bees. There are four British species, all of them rare or very scarce. The pattern of veins in the forewing is distinctive. As in other genera of Megach-ilidae there are two submarginal cells, but the second outer cross-vein meets the median vein further out than does the first outer-cross vein (close to the same point or towards the wing-base in other genera) (Fig. 2.7). *Stelis breviuscula* (Fig. 2.63) is very similar to its host, *Heriades truncorum*, but lacks the pollen-hairs in the female. Like its host it has a very limited distribution in south-eastern England, but appears to be expanding its range quite rapidly. *Stelis punctulatissima* takes *Anthidium manicatum* as one of its two hosts, and so can be searched for where that species is common. However, it is much less common than its host, and most records are from southern or south-eastern England.

Family Apidae

This very large and diverse family includes the bumblebees (*Bombus* species) and honeybees (*Apis*), which are beyond the scope of this handbook. The genera that concern us are a mixture of cleptoparasites (three genera) and four genera of solitary species, which were formerly included in the family Anthophoridae (or subfamily Anthophorinae). Except in the cleptoparasites, the females have pollen-hairs on the hind legs.

Genus *Eucera*

Two species have been recorded in the British Isles, but one has not been noted for more than 50 years and is believed to be extinct. The other, *Eucera longicornis* (Fig. 2.64), is so-called because of the exceptionally long antennae in the male. The male has long, white plumose hairs on the face and underside, those on the thorax grading to pale fawn-brown on the dorsal surface. The clypeus is bright yellow, and there are dense fringes of white hairs on abdominal tergites 4 and 5. The female is a chunky bee, with antennae of normal length. The colour pattern is similar to that of the male, but with a dense incomplete band of flattened white hair on the rear margin of the fourth abdominal tergite, and dense golden hair on the fifth and sixth tergites. The head is black, and the tongue is very long. The pattern of venation in the forewing is similar to that of *Megachile* (two submarginal cells, with the second outer cross-vein meeting the median vein closer to the wing base than does the first outer cross-vein), but the basal vein is strongly arched, as in *Halictus* and *Lasioglossum*) (Fig. 2.8). The bee is very local across southern England and south Wales, and is probably in decline. It nests in aggregations in burrows in open woodland and on coastal grasslands with eroding cliffs or sea walls. It is said to collect pollen from a range of flowers in the pea family (Fabaceae), and in southern Britain seems to be very closely associated with meadow vetchling (*Lathyrus pratensis*). It flies from late May until July.

Genus *Anthophora*

There are five species that occur in the British Isles. They are medium to large, long-tongued bees, with a compact, robust appearance. According to species, the females make their nests in a wide variety of holes and crevices, both above ground (in rotting wood, old walls or even garden furniture) or below ground in burrows, which they dig. The wing venation is distinctive: three submarginal cells, with the second approximately equal in length to the third, and with the first and second outer cross-veins meeting the median vein at the same point (Fig. 2.9). Males of all species have areas of yellow cuticle on the face, and in both sexes the lower part of the face is prominent and the labrum fits into a shallow U-shaped concavity in the clypeus (Fig. 2.20).

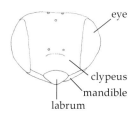

Fig. 2.20 Face of *Anthophora*.

The hairy-footed flower bee, *Anthophora plumipes* (Fig. 2.65). This species is one of the most familiar bees in parks and gardens. The males and females are very different and might be taken by the uninitiated as two distinct species. The males have a dense covering of golden-brown hair that gives way to black on the rear tergites of the abdomen, but the hair colour soon fades, and old individuals may be pale grey-white. There are extensive yellow areas on the face, and a yellow spot near the base of each mandible. There is a graceful drape of long hairs arising from the feet of the mid legs. They establish patrol routes in early spring, well before the first arrival of the females. The latter are covered in dense black hairs, but with bright orange pollen-hairs on the hind tibiae and basitarsi. These could easily be mistaken for full pollen-baskets of a bumblebee – and both sexes have the general appearance of small bumblebees. They nest, often in aggregations, in holes and crevices in old walls, or in mud banks and coastal cliffs. The species is widespread and often common in southern England, becoming more localised further north to Yorkshire.

Anthophora bimaculata (Fig. 2.66). In the British Isles this species is confined to southern England and East Anglia, where it is often abundant on the sandy soils of heaths and coastal dunes, and where they nest in burrows. Both sexes are smaller than the previous species, have blue-green eyes, and both sexes have distinctively-shaped yellow areas on the face and mandibles. The females have white pollen-hairs on the hind tibiae and basitarsi, and the males have narrow white hair fringes on the rear of the abdominal tergites. They fly later than *A. plumipes*, from late June to early September.

Genus *Ceratina*

There is just one species in the British Isles, the small carpenter bee, *Ceratina cyanea* (Fig. 2.67). This is a small sparsely haired bee with beautiful metallic blue-green reflections. It makes its nest in hollowed-out stems of bramble or holes in wood, hence the term carpenter bee. Its distribution in the British Isles seems to be restricted to the south-eastern counties of England, extending into East Anglia, but currently it seems to be expanding its range. They have a long flight season, and can be seen from early spring through to September. Both sexes hibernate as adults, in plant stems. The male has white markings on the face, and both sexes have predominantly black legs. Females do not have obvious pollen hairs. The short, blunt mandibles, tipped by three points, are a distinctive feature. The forewing venation is similar to that of *Andrena* (Fig. 2.5).

Genus *Xylocopa*

One species, the violet carpenter bee (*Xylocopa violacea*), is sometimes recorded in Britain, but so far has not established itself as a resident species. Holiday-making bee-watchers in southern Europe will be familiar with this species, or one of its close relatives. They are large bees (20 mm or more in length) with purple-tinted wings and a dense coat of purple-black hair. They nest in a variety of holes and crevices and also can make their nest-holes by digging into dead wood with their powerful mandibles. There are reports of attempted breeding in southern Britain, and it seems possible that this distinctive bee will soon become a resident species.

Genus *Nomada*

This is a very large genus, comprising some 34 species in the British Isles (including five that occur in the Channel Islands only). All species are cleptoparasites, mostly invading the nests of *Andrena* species. Several species have a pattern of black-and-yellow bands on the abdomen with yellow markings on the thorax, giving the appearance of very small wasps. Others are red-brown with yellow bands or patches, while still others have a range of black and red patterns. The body, especially the abdomen, is very sparsely haired. The wing venation is similar to that in *Andrena*, except that the third submarginal cell is the same length as, or shorter than, the second (longer in *Andrena*) (Fig. 2.10). There are patches or bands of yellow on the legs and (in males) yellow or reddish markings on the face and/or mandibles. Most species cannot be definitively identified without microscopic examination.

Nomada goodeniana (Fig. 2.68). This is one of the more common and widespread species in the genus. It is wasp-like, with yellow bands across tergites 1 to 5, that in tergite 1 being narrowly interrupted in the middle. The dorsal surface of the thorax is black, with two yellow-marked tubercles on the scutellum, and yellow on the pronotum and tegulae (Fig. 2.21). Females have orange-red antennae, while the males have the first seven segments of the antennae black dorsally. It invades the nests of large *Andrena* species, and flies in spring, with a much scarcer second brood in August.

Nomada marshamella (Fig. 2.69). This is another common and widespread wasp-like species. It is very similar to the previous species, but the tegulae are orange, never yellow, and the yellow bands on tergites 2 and 3 are narrowly interrupted in the middle. It uses the nests of *Andrena scotica*, and possibly other large *Andrena* species.

Nomada fucata (Fig. 2.70). This is another wasp look-alike. It has tubercles, pronotum and tegulae all yellow-marked (like *N. goodeniana*), but the yellow on the scutellum is fused as a single yellow patch. Tergite 1 is brown at the base (no yellow band), and the yellow band in tergite 2 is interrupted in the middle. This species is quite common in southern and south-eastern England and south Wales, following the distribution of *Andrena flavipes*, its main host. Both species seem to be expanding their range.

Nomada flava and *Nomada panzeri* (Fig. 2.71). These species are very difficult to distinguish from each other, but *N. flava* is a southern species, while *N. panzeri* is widely distributed across the British Isles, including Scotland and Ireland. The females have longitudinal red stripes on the mesonotum, and the tubercles and tegulae are red (not yellow). Tergite 1 of the abdomen is black basally and then red, and there are extensive yellow markings on the rest of the abdominal tergites. The hosts of these species are several of the larger *Andrena* species.

Nomada fabriciana (Fig. 2.72). This species has yet another colour pattern: predominantly red on the abdominal tergites

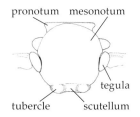

Fig. 2.21 Thorax of *Nomada*, viewed from above.

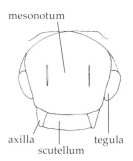

mesonotum

axilla   tegula
    scutellum

Fig. 2.22 Thorax of
Epeolus, viewed from
above.

♂ male
♀ female

(with yellow spots on tergite 2, or 2 and 3), and the dorsal surfaces of the thorax and the head are all black. The female is distinctive in having segments 8 to 11 of the antennae black, the rest orange-red. Several *Andrena* species have been recorded as hosts. It is widely distributed in the British Isles, including Scotland and Ireland, but it is probably more common in southern and south-eastern England.

Genus *Epeolus*

There are just two species in the British Isles, both cleptoparasites of various *Colletes* species. They are almost hairless except for patches of fine white or yellowish flattened hairs. These are formed into paired spots on the abdominal tergites. There are orange areas on the legs in both species, and backward-pointing plates (axillae) on either side of the scutellum (Fig. 2.22). The forewing venation is similar to that of *Nomada* (Fig. 2.10). *Epeolus cruciger* (Fig. 2.73) invades the nests of *Colletes succinctus* as well as the more localised *Colletes marginatus*. It is widely distributed where its hosts occur, in southern England and Wales, northwards as far as north Yorkshire. *Epeolus variegatus* (Fig. 2.74) is a nest parasite of several *Colletes* species, including *C. fodiens* and *C. halophilus*. The flight periods of both species coincide with those of their hosts.

Genus *Melecta*

There were two species in the British Isles, both cleptoparasites of *Anthophora* species. However, one is believed to have become extinct, along with the steep decline of its very scarce host. The survivor, *Melecta albifrons* (Fig. 2.75), whose host is *Anthophora plumipes*, is widely distributed in gardens and coastal sites in southern and south-eastern England and South Wales, with a scattering of more northerly reports. Like its host it is a spring species, flying from April to early June. The cuticle is black, with sparse long hair on the abdomen, and patches of white or pale grey-brown hair on the thorax, legs and head. There are also distinctive paired patches of white hairs on abdominal tergites 1 to 4 or 5. There are darker forms in which white is replaced by dull yellow-brown, or the pale patches may be altogether missing. The forewing venation is similar to that of *Nomada* and *Epeolus*, assuming that the length of the third submarginal cell is measured outwards to its intersection with the second outer cross-vein (Fig. 2.10). Because the third submarginal cell bulges out beyond that, the venation has more the general appearance of Fig. 2.5.

Fig. 2.23 *Colletes succinctus* ♀

Fig. 2.24 *Colletes halophilus*, mating pair

Fig. 2.25 *Colletes hederae* ♀

Fig. 2.26 *Colletes cunicularius* ♀

Fig. 2.27 *Hylaeus communis* ♀

Fig. 2.28 *Hylaeus pectoralis* ♂

Fig. 2.29 *Hylaeus hyalinatus* ♀

Fig. 2.30 *Hylaeus hyalinatus* ♂

Fig. 2.31 *Andrena fulva* ♀
(tawny mining bee)

Fig. 2.32 *Andrena cineraria* ♀

Fig. 2.33 *Andrena haemorrhoa* ♀

Fig. 2.34 *Andrena clarkella* ♀

Fig. 2.35 *Andrena florea* ♀

Fig. 2.36 *Andrena hattorfiana* ♀

Fig. 2.37 *Andrena marginata* ♀

Fig. 2.38 *Halictus rubicundus* ♀

Fig. 2.39 *Halictus tumulorum* ♀

Fig. 2.40 *Lasioglossum calceatum* ♂

Fig. 2.41 *Lasioglossum malachurum* ♀

Fig. 2.42 *Lasioglossum morio* ♀

Fig. 2.43 *Sphecodes ephippius*

Fig. 2.44 *Sphecodes monilicornis*

Fig. 2.45 *Sphecodes puncticeps*

Fig. 2.46 *Macropis europaea* ♀

Fig. 2.47 *Dasypoda hirtipes* ♀

Fig. 2.48 *Melitta haemorrhoidalis* ♀

Fig. 2.49 *Melitta leporina* ♂

Fig. 2.50 *Melitta tricincta* ♂

Fig. 2.51 *Anthidium manicatum* ♂ (wool-carder bee)

Fig. 2.52 *Megachile willughbiella* ♂

Fig. 2.53 *Megachile centuncularis* ♂

Fig. 2.54 *Megachile leachella* ♀

Fig. 2.55 *Osmia bicornis* ♀ (red mason bee)

Fig. 2.56 *Osmia caerulescens* ♂

Fig. 2.57 *Osmia leaiana* ♀

Fig. 2.58 *Osmia bicolor* ♀

Fig. 2.59 *Osmia spinulosa* ♀

Fig. 2.60 *Heriades truncorum* ♀

Fig. 2.61 *Chelostoma campanularum* ♀

Fig. 2.62 *Coelioxys elongata* ♀

Fig. 2.63 *Stelis breviuscula* ♂

Fig. 2.64 *Eucera longicornis* ♂ (long-horned bee)

Fig. 2.65 *Anthophora plumipes* ♀ (hairy-footed flower bee)

Fig. 2.66 *Anthophora bimaculata* ♀

Fig. 2.67 *Ceratina cyanea* ♂ (small carpenter bee)

Fig. 2.68 *Nomada goodeniana* ♂

Fig. 2.69 *Nomada marshamella* ♀

Fig. 2.70 *Nomada fucata* ♀

Fig. 2.71 *Nomada flava* or *panzeri* ♂

Fig. 2.72 *Nomada fabriciana* ♂

Fig. 2.73 *Epeolus cruciger*

Fig. 2.74 *Epeolus variegatus* ♂

Fig. 2.75 *Melecta albifrons* ♂

3 Bee lives

3.1 The life cycle

The great majority of insects lay their eggs in, on, or close to the potential food source of their offspring. Butterfly eggs are usually laid on carefully selected larval host plants, but otherwise there is no further concern for their fate. Most damselflies and dragonflies lay their eggs in plant tissue or open water, leaving their offspring to fend for themselves. The social bees, ants and wasps go much further in constructing nests and providing a cohort of workers to nurture their offspring. The solitary bees can be divided into two groups. The majority make their own nests, usually consisting of a series of brood cells, each of which is primed with a food mass. The female lays an egg in each cell, then closes it and proceeds to construct another. However, a surprising number of species (belonging to five genera in the British Isles) lay their eggs in the cells of other bees. The resulting larvae consume the provisions supplied by the host female. As there are many distinctive features of these cleptoparasitic species, and they are of interest in their own right, they will have a chapter to themselves (Chapter 4). The rest of this chapter will focus on the majority of solitary bees, the ones that make and provision their own nests.

The eggs are elongated and gently curved, and are laid singly in the cell, either on the surface of the food mass, or attached to the cell wall. If the egg is viable, a tiny embryo develops within the outer casing. The life-cycle in bees has a complete metamorphosis, meaning that there are four very different developmental stages: egg, larva, pupa and adult. In this respect bees resemble insect groups such as flies, butterflies and beetles, and are unlike groups such as dragonflies, grasshoppers and crickets whose immature stages resemble the adults in external appearance.

There are five larval stages (instars) in bees, but in many species the first stage is completed while the insect is still within the egg, so that the observable instars are reduced to four. Initially the larva is grub- or maggot-like, having no legs and very little external structure. It is adapted solely to consuming the mass of food provided for it. As it does so it casts its cuticle several times. For the first three instars the young larva has no connection between the front and rear parts of its gut, and so is not able to defaecate. This could be an adaptation that reduces the likelihood of fungal or

cuticle
the tough outer layer on the surface of the body

Fig. 3.1 Brood cells of *Osmia leaiana*, showing cocoons, and nest-plug at the right hand end.

Fig. 3.2 Brood cells of *Osmia leaiana* with cocoon opened to show final instar larva.

bacterial infections within the cell. The connection is made when the larva has completed its feeding phase, and at the final instar it is able to defaecate. At this stage the larva has a tougher external cuticle and a firmer, more distinctive shape. This is known as the prepupal stage.

At this point there are two possibilities for further development. In temperate climates such as ours, living beings have to evolve ways of passing seasons with inclement weather and where opportunities for finding food are in short supply. For most species, passing the winter months presents a challenge, and while some continue to be active, or simply migrate to warmer climates, insects, including solitary bees, generally enter into a dormant phase. Among the bees the most frequent stage for passing the winter months is as a prepupa, still within its brood cell (Figs. 3.1–3.2). In these species, the next, pupal, stage is reached the following spring and the adults emerge later in the spring, or during the summer. In some species a proportion of the prepupae continue in that stage over two (or even more) winters. This is possibly a reproductive strategy that spreads the risks associated with high rates of mortality brought about by exceptionally bad weather in any one year (Michener, 2007).

Alternatively, the prepupa may go on directly to shed its cuticle for the last time, revealing that it has reached the next stage in the life-cycle: the pupa. In most bees the pupa simply relies on the brood cell for continued protection, but in the family Megachilidae the fully developed larva spins a silken cocoon before pupation. This stage in the life-cycle is sometimes referred to as resting, because the pupa has no means of locomotion and does not feed. However, this is misleading, as it is the stage in the insect's life-history during which the most remarkable transformation takes place. The internal structures are reorganised and the quite different body-plan of the adult bee is formed within the outer casing. Externally, the pupa displays all the features of the future adult, with the mouthparts extruded and the wings represented by small flaps on each side. As the internal transformations take place, the colouring of the pupa becomes darker, often with the eyes clearly defined. When the reorganisation is complete, the pupal cuticle is broken open and the fresh adult emerges. Blood is then pumped through the wing-veins and the wings assume their final shape. Now the adult bee may break out of its cell, or it may remain as an adult within the cell, as an alternative means of passing the winter.

One species, *Ceratina cyanea* (the small carpenter bee), illustrates another variation on the timing of the stages in its life-cycle. Mating and egg-laying take place in May, with a fresh generation emerging during August and September. These adults may be seen at flowers at that time, but go on to spend the autumn and winter hibernating in hollowed-out plant stems (especially bramble). It is even suggested that some adults may pass two winters, giving them an adult life-span of as long as 18 months (Else, 1995) (Fig. 3.3).

Bees belonging to some species in the genera *Lasioglossum* and *Halictus* also fly during the summer and autumn, with only the previously mated females surviving to hibernate over winter. These establish nests in spring, and rear a first generation of non-reproductive workers. These help with making and provisioning a further series of brood cells from which emerge males and fertile females (gynes) during the summer. There is evidence that in some of these species a female may lay her eggs over two successive seasons (Field, 1996). Species that over-winter as adults often emerge very early in the following spring, as they have no further developmental stages to complete. This is a feature of the life-cycle in many mining bees of the genus *Andrena*. These remain as adults in their brood cells over winter and are often the first species to be seen in spring – often as early as March. Some species, such as

Fig. 3.3 *Ceratina cyanea* (small carpenter bee), male.

Fig. 3.4 *Andrena flavipes*, summer brood.

mining bees *Andrena flavipes*, *Andrena minutula* and *Andrena bicolor* manage to complete two full life-cycles in a single year. The spring and summer broods of double-brooded species often differ in appearance (Fig. 3.4).

The adult bees, once they have emerged from their natal nests, have three main priorities: finding a mate, feeding, and, in the case of females, finding or making a place to nest, so beginning a new reproductive cycle.

3.2 Mating systems: reproduction in bees

In most sexually reproducing animals what is called the mating system consists of a series of stages. These include: *rapprochement*, the coming into contact of male and female; *courtship*, activities, usually performed by the male, to induce receptivity in the female; *copulation*, genital engagement between the two; *insemination*, the passing of sperm into the female; *fertilisation*, fusion of egg and sperm; and subsequent phases, not common among insects, in which one or both parents invest in the care or nurturing of their offspring. In bees, as with other Hymenoptera, males result from unfertilised eggs, so females determine the sex of their offspring as they lay their eggs. Males have an additional segment to the antennae (13, as against 12 in the female), and the antennae are longer in the males – sometimes, as in the long-horned bees, genus *Eucera*, very much longer (Fig. 2.64). There is evidence that chemical signals, emitted by females and detected by the male antennae, play an important role in several phases of the mating system. The rapprochement phase involves several searching strategies on the part of males, but among the bees that occur in the British Isles there appears to be little or nothing that could count as courtship prior to the male grasping the female. However, even when a male has a female in his grasp, mating does not follow automatically. In some species (notably some *Colletes* and *Lasioglossum* species), several males may grasp the same female, and a scramble ensues in which only one of them gains genital contact. In many other species, a male may remain on top of the female for some time, often stroking her antennae with his own, before actual copulation takes place. Copulation and insemination take place at the same time, but sperm are stored internally by the female and fertilisation takes place later, when she lays each egg. Males of species that occur in the British Isles are not known to play any part at all in the nurturing of their offspring beyond their role in fertilisation. Females (except in the case of the cleptoparasites), however, are unlike the vast majority of insects in the preparations they make for the well-being of their offspring: finding or making a nest, stocking each brood cell with sufficient food, with the right mix of nutrients to see the resulting larva through the whole of its development, sealing and, in some cases, concealing the nest. In the case of the solitary species, there is no subsequent care of the offspring, although there is evidence that females of *Ceratina* may sometimes remain within the nest burrow and guard the brood cells until after the fresh adults emerge (Else, 1995). In addition to the bumblebees and honeybees, which are well-known for continuous care of the offspring by a distinct worker caste, there are also some species belonging to other genera that show varying degrees of sociality. The small black halictine bee, *Lasioglossum malachurum* (also known as *Lasioglossum malachura*), is perhaps the best known of these.

Finding a mate

According to species, male bees have a variety of methods for finding potential mates. They usually emerge earlier than the females (protandry) and spend several days familiarising themselves with their surroundings. Many mining bees nest in dense aggregations. In these species, males frequently remain close to their natal nest site, and establish patrolling routes, frequently stopping to investigate burrow entrances. On the Essex coast, numerous males of *Megachile leachella* were observed flying fast and very close to the ground on sparsely vegetated sand dunes, stopping at nest entrances, and also contouring sea holly plants, diving at

Fig. 3.5 Patrolling habitat of male *Megachile leachella* on the Essex coast with inset of male *M. leachella*.

Fig. 3.6 Male *Osmia bicornis*.

any other insects – including Essex skipper butterflies – foraging on the flowers (own observation) (Fig. 3.5). At another site, however, males of the same species were observed settling on raised vegetation, darting off occasionally to buzz another passing male, following it for 2–3 metres and then returning to the original vantage point. These observations suggest that males of a single species may have more than one strategy in their repertoire, and also that patrolling behaviour involves both male-to-male competition as well as mate-searching.

As females of most solitary bees are believed to mate only once, it is not surprising that competition between males is intense. Protandry may be one adaptation to this – advantages go to the males that are on site and ready to mate when the first females emerge. But there is a risk if the males emerge too early that they will not survive to benefit. In 2014, males of *Osmia bicornis* (formerly known as *Osmia rufa*), the red mason bee, were first seen in my back garden more than a month before the first females (Fig. 3.6). In several species of the ground-nesting *Colletes* bees, males actually dig down to intercept newly emerging females even before they reach the surface (O'Toole, 2013). In one species, *Colletes cunicularius*, whose mating behaviour has been studied in detail, the searching activity and direction-finding of the males is guided by both chemical and vibratory cues emitted by virgin females as they emerge from their brood cells (Fig. 3.7). The pheromone concerned, linalool, is secreted by the mandibular glands of the female and patrolling males detect the signal by way of receptors in their antennae. Vibratory communication by buzzing

pheromone

a substance produced by an animal that influences the behaviour or physiological development of another individual of the same species

Fig. 3.7 Two males of *Colletes cunicularius*, one of them digging to reach a newly emerged female.

et al.
is short for *et alia* (and
others)

Fig. 3.8 (a) Males of *Colletes halophilus* cluster around a female, (b) mating pair of *Colletes cunicularius.*

complements chemical signals in guiding the direction of searching by the males. But males also buzz during copulation, possibly increasing female receptivity (Cane & Tengö, 1981; Larsen *et al.*, 1986; Borg-Karlson *et al.*, 2003). Once at the surface, the female will often be pursued by up to ten males, which form a wriggling mass around her, each male struggling for position. Eventually one of them succeeds and the rest soon disperse to resume their patrol (Fig. 3.8). Males of *Lasioglossum malachurum* exhibit similar behaviour, flying in large numbers low over nesting aggregations and attempting to mate with females as they emerge – usually several males clustering round a female. In this species, too, chemical signals emitted by the females are involved in attracting males. The constituents of the sex attractant pheromones in this case are hydrocarbons and isopentenyl esters, which have low volatility and so would be effective over only short distances (Ayasse *et al.*, 1993; Ayasse *et al.*, 1999). Females of the mining bee *Andrena nigroaenea* also emit a sexual attractant which is mimicked by the early spider orchid as a means of attracting males to trigger their pollination mechanism (see Chapter 5) (Fig. 3.9).

Fig. 3.9 Male *Andrena nigroaenea.*

Fig. 3.10 A male *Colletes halophilus* approaches a returning female.

In the *Colletes* bees mentioned above, males pay very little attention to females that have already mated. It may be that females emit a scent-signal that indicates that they are not receptive (O'Toole, 2013). At a nest aggregation of *Colletes halophilus* on the Essex coast, males were observed to react very differently to females with pollen loads returning to their nests than to freshly emerging females. These were often briefly approached (Fig. 3.10), but the males quickly departed (own observation). In species whose females mate only once, females may benefit from signalling to males that they have already mated, since, for example, there is evidence that their foraging efficiency can be significantly reduced by male harassment. But where females mate more than once, a male may benefit from either ensuring that his sperm fertilise the female's eggs, or reducing the likelihood that she will mate with another male. One way of doing this is for the male to secrete and pass on to the female an antiaphrodisiac pheromone. In female *L. malachurum*, receptive females produce more volatiles than non-receptive ones, and there are also differences in the proportions between the constituents (Hefetz, 1998; Smith & Weller, 1989). In *Andrena nigroaenea* another pheromone (farnesyl hexanoate) is produced by females after mating, and has the effect of inhibiting male mating behaviour (Schiestl & Ayasse, 2000). In *Osmia bicornis*, too, males are able to discriminate between virgin and previously mated females by scent (Dutzler & Ayasse, 1996). Evidence of males transferring antiaphrodisiac pheromones to females has been less readily available, but it has been shown in *L. malachurum* (Ayasse, 1994) and also in *Osmia bicornis*, in which the males rub a product of their sternal glands onto the wings of females after mating (Ayasse & Dutzler, 1998).

Osmia bicornis is a familiar species in gardens, and readily nests in artificial bee hotels. Males of this species gather around such nest sites as females are emerging, and competitive behaviour among them leads to the establishment of dominance hierarchies (Raw, 1972). Several other members of the *Osmia* genus make their nests in snail shells, and in these species the males spend much of their time exploring in and around snail shells, whether waiting for virgin females to emerge, or establishing a territory at an empty shell, as in *Osmia aurulenta* (O'Toole, 2013). Males of *Osmia spinulosa* fly low from one snail shell to another, stopping to investigate each one (Fig. 3.11). An alternative strategy for locating a potential mate is to search for females at patches of flowers where they forage. Males of the small black bee, *Macropis europaea,* search for females at flowers of yellow loosestrife. The strategy works, as the females specialise in foraging from this plant. Males, often dozens at any one time, fly circuits through and around patches of the plant, diving at any bee they see foraging, including other males (Fig. 3.12). When a male succeeds in pouncing on a female, the pair may drop to the ground, or, presumably when the male's advances are rejected, there is a brief struggle before the male departs and the female continues

Fig. 3.11 Male of *Osmia spinulosa* waiting at a snail shell.

Fig. 3.12 A 'face-off' between two male *Macropis europaea*.

Fig. 3.13 A failed mating-attempt by a male *Macropis europaea*.

Fig. 3.14 Contrasting yellow face of male *Macropis europaea*.

foraging (Fig. 3.13). In this and numerous other species in which the males search for females at flowers, males have bright, pale faces, contrasting with the rest of their body and hair (Fig. 3.14). This may have a role in species recognition, or in signalling between rival males.

In the case of the hairy-footed flower bee, *Anthophora plumipes*, some of whose males adopt a similar strategy, there is substantial evidence of the loss of foraging efficiency on the part of the females as a result of male harassment. The male activity is so intense, the foraging of the females appears to be significantly interrupted. One study (Stone, 1995) showed that the most active males at their daily peak made as many as eleven mating attempts per minute. This disruption reduced by a half the foraging success of the females for nectar. A notable response to high density of males at a patch of comfrey was that the females shifted their foraging to flowers within the patch while males tended to focus their attention on outer flowers. Experimental removal of the males resulted in the females shifting back to forage on the outer flowers, which could be done more efficiently in the absence of male harassment.

Yet another male strategy, exemplified by some males of the common wool-carder bee (*Anthidium manicatum*), is to occupy and defend a territory. In this case the territory is usually a patch of flowering plants, or a length of hedgerow, encompassing plants that are used by females as sources of pollen, nectar or nesting materials – or all three. Literature descriptions relate to territories used for all three resources, but in many cases the territories include floral resources only, and females search for nesting materials elsewhere (see below). Territorial males spend much of their time

Fig. 3.15 A female *Anthidium manicatum* forages while a male patrols below.

Fig. 3.16 A mating pair of *Anthidium manicatum*.

polygyny
a mating system in which a male mates with more than one female

patrolling their territory, with short bursts of swift, direct flight, interrupted by moments of hovering before abrupt changes of direction. These males hover, seeming to peer attentively at a flower, or flower-cluster, as part of their search strategy (Fig. 3.15). Other insects, including other males, but also flies, other bee species, and even butterflies are treated aggressively. Usually these interactions consist of the territorial *Anthidium* turning towards the intruder and approaching it, or following on if it retreats. Any other insect spotted foraging will be dived at and driven off. The males have sharp spines on the rear segments of the abdomen which they can use to inflict damage by holding a victim between their legs and bending the rear part of the body forwards. Severinghaus *et al.* (1981) report *Anthidium* males wounding and disabling numerous honeybees. Conflict with other *Anthidium* males that trespass on an established territory usually takes the form of a visual 'stand-off', the intruder leaving without a fight, but males in control of a territory make periodic excursions elsewhere, especially when floral resources in the home territory decline. This results in contests between rival males for territories, but these are usually conducted by spiralling flights rather than by direct physical combat. According to Severinghaus *et al.* (1981) it is usually the larger male that succeeds in these confrontations, although old and relatively weak males can also lose out to smaller males.

The tendency for territorial males to attack and chase off intruders of other species can be interpreted as a means of maintaining high levels of floral resources in the territory, making it more attractive to foraging females than would otherwise be the case. When a female enters a territory and is observed by the incumbent male, he approaches and follows her if she is in flight. When she settles and begins foraging from a flower he pounces on her and usually succeeds in mating. O'Toole (2013) reports a territorial male observed in his Oxford garden as mating as often as 16 times in one day. It seems that this form of resource-based polygyny mating system is associated with females accepting frequent copulations, and it may have resulted in sexual selection for larger males (Fig. 3.16). Though there is considerable size variation among the males, they are on average significantly larger than the females. In most solitary bees, the males are considerably smaller than the females.

Not all males are able to hold territories, and those that do not, usually smaller ones, have an alternative strategy. They forage over a wider range, enter discreetly into the territories of other males and attempt to mate with foraging females when the incumbent male is absent or in a different part of the territory. These satellite males achieve fewer copulations than territorial males because they encounter fewer females (and sometimes their copulations are actually interrupted by attacks from the larger male), though their mating attempts are not rejected by females more frequently than those of the territory owner. It seems that this is not a fixed behaviour for any individual, as former satellite males can take over a territory at another time, depending on the competitive pressure among the males.

The yellow-faced males of some species of the genus *Hylaeus* are territorial around flowers visited by females. A study of Australian species (Alcock & Houston, 1996) compared males of territorial species with non-territorial ones. The former tended to be larger than the females of their species, and had anatomical modifications in the form of abdominal projections or spines used in grappling with rivals. The males of non-territorial species were not larger than the females, and lacked armaments. The males of two species found in the British Isles, *Hylaeus signatus* and *Hylaeus confusus* have prominent projections from the underside of the third abdominal segment (Fig. 3.17).

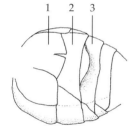

Fig. 3.17 View of the underside of the abdomen of a male *Hylaeus signatus*, showing semi-circular projection on tergite 3.

Males of the hairy-footed flower bee, *Anthophora plumipes*, adopt territories when the density of males at an area of flowering shrubs is low, but as density increases (especially later in their flight period), they revert to flying in overlapping patrol routes. Stone *et al.* (1995) consider that at low density males may be able to defend a resource patch and monopolise mating opportunities with relatively little cost, but with greater density of competing males, defence of a territory would require greater expenditure of energy and impose other costs, prompting a shift to patrolling and opportunistic mating attempts. In this species mating may take as long as 40 minutes, during which time a territorial male would, in any case, not be able to prevent intruders from mating with any females that arrived.

These male strategies for finding receptive females exploit the females' foraging preferences, and it is because of this that the mating systems are referred to as resource-based. However, there is yet another approach, greatly dependent on chemical communication. Males establish patrolling routes and mark stopping-points along them with pheromones which are attractive both to other males of the same species and to females. Darwin first studied this behaviour on the part of bumblebees in his garden, although he did not know that pheromones were involved (Benton, 2006). Similar patterns of male behaviour are also found in some groups of solitary bees. Tengö and fellow researchers have studied this non-resource based patrolling behaviour in species of *Andrena*, and have identified the lemon-smelling pheromones that are used. These compounds are secreted by the males from mandibular and labial glands in the head, and are distinctive for each species. This has been seen as a means of species isolation, but more recently the individual variation in the chemical blends emitted has led to their interpretation in terms of kin recognition and sexual selection (Tengö & Bergström, 1976, 1977; Tengö, 1979; Tengö *et al.*, 1990; Ayasse *et al.*, 2001) (Fig. 3.18).

In most species of solitary bee, there is no advanced courtship behaviour, such as is often seen in other insect groups (notably among grasshoppers and crickets, see chapters 4–6 of Benton, 2012). Patrolling or territorial males generally dart at any female they see and attempt to mount her. This has the appearance of what is sometimes called coercive copulation, in which female preference plays no part. However, as noted above, males do not always attempt to mate, presumably as a result of some visual or chemical signal communicating that the female is not receptive. Even after a male has succeeded in momentarily mounting a female, she may throw him off, or suddenly drop to the ground.

Fig. 3.18 Male *Melecta albifrons*, apparently scent-marking vegetation on its patrol route.

Males of *Anthophora plumipes* approach and follow females in flight, taking up a position 5–8 cm behind their quarry. A line of three or four males may join the queue, each flying a few centimetres behind the one ahead. In the study reported by Stone *et al.* (1995) the male in the lead position was the one most likely to successfully mount the female when she landed on a flower. Other males sometimes challenged the male in pole position by chasing, hovering face-to-face, clashing or grappling. Male size was significantly positively correlated with success in these conflicts. In this as in other species, male size confers two advantages. The first is the greater ability of larger males to defend territories, or to maintain lead position in following females in non-territorial situations. The second has to do with thermoregulation. As an early spring species, *Anthophora plumipes* encounters low temperatures and the bees require more-or-less extended periods of warm-up before the temperature of their thorax, which contains the wing muscles, reaches the point at which flight is possible. In early morning, as ambient temperatures rise, males approach the mouth of the tunnels where they spend the night. On sunny days they bask, thus saving energy that would otherwise be spent on the rapid muscular contractions that are used to raise thoracic temperatures (endothermy). Both males and females begin their daily activity with foraging for nectar, before turning, in the case of males, to mate-searching, or, in the case of females, to establishing and priming nests. Because larger bees have smaller surface area to volume ratios they tend to be able to sustain flight at lower ambient temperatures than smaller ones. This would give larger males the ability to conduct their mate-searching activities over longer periods during the day, and also to be active during more adverse weather conditions (Stone *et al.*, 1995; Stone & Wilmer, 1989; Stone, 1993).

In some species there may be a considerable time lapse between the male mounting the female and actual copulation. Possibly because of the ease with which it can be attracted to occupy artificial nests, the mating behaviour of the red mason bee, *Osmia bicornis*, has been closely studied. Once a male has mounted a female there is an extended period of some 10 minutes or more of precopulatory behaviour before actual mating takes place (Siedelmann, 1999). During this period the male strokes the antennae of the female with his own, places his front legs over her eyes, and emits bursts of a high-pitched humming sound, which is produced by thoracic vibrations. He then moves backwards on the female and attempts to insert his genitalia into her genital chamber. At this point she may struggle to shake him off, or turn her abdomen down to avoid genital contact. However, if the male's attempt is accepted copulation persists for several minutes, after which the male stays in place and strokes the dorsal surface of her abdomen with the ventral surface of his own for several more minutes. This postcopulatory behaviour involves the application of an antiaphrodisiac pheromone which reduces the appeal of the female to subsequent potential mates. This sequence shows that female choice is possible, and a study by Conrad *et al.* (2010) investigated the criteria involved in expressions of female preference in this species. In this study, females were significantly more likely to allow mating with males that produced longer bursts of thoracic vibration than others. They also preferentially mated with males slightly larger than average (but not the largest males), and with more closely related males. Chemical analysis of cuticular pheromones produced equivocal evidence, but did suggest that females could assess closeness of relationship between

Fig. 3.19 A precopulatory pair of *Osmia leaiana*.

themselves and particular males, and also appeared to show definite preferences for specific pheromonal bouquets (Fig. 3.19).

In our own garden, a male *Melecta albifrons* (a cleptoparasite of the hairy-footed flower bee, *Anthophora plumipes*) was observed to fly at a foraging female, hitting her and flying on, only to return for another hit. This time the female fell to the ground, followed by the male, who immediately mounted her. After a brief period of leg-twitching and re-positioning, the male grasped the female around the waist with his hind legs, curved his mid tarsi round the leading edge of her forewings, preventing flight, and placed his fore legs each side of her neck. For several minutes copulation continued, with the female's antennae pointing directly up to the face of the male, while he repeatedly raised and lowered his antennae so that they stroked the flagella of her upright antennae (Fig. 3.20).

The mating procedure of another frequent garden visitor is still more complex. The leaf-cutter bee *Megachile willughbiella* is one of several species of *Megachile* in which the front tarsi of the male form wide, externally flattened pads, from which are draped fans of long hairs, or setae. When the male mounts a female, he positions himself so that the fans of hair cover her eyes. Puzzled by this, Wittmann & Blochtein (1995) observed in detail the anatomical adaptations and related mating behaviour in this species. The male holds the head of the female in place with prominent, forward-pointed spines on the fore coxae (Fig. 2.11), holds down her wings with the middle legs, and grips her abdomen with his hind legs, raising it up. Meanwhile, her antennae

flagellum (plural flagella) the outer 9 (females) or 10 (males) segments of the antenna

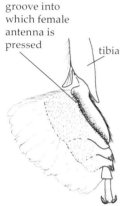

groove into
which female
antenna is
pressed   tibia

Fig. 3.21 Inner surface of
the fore tarsus of a male
Megachile willughbiella.

Fig. 3.20 A mating pair of *Melecta albifrons*. The female antennae
are held erect, while the male strokes them with his antennae.

are caught between his gaping mandibles, and part of the
flagella pressed into deep, elongated grooves on the inner
surface of the greatly expanded first tarsal segments of the
fore legs (Fig. 3.21). A result of this is that the drapes of
hairs attached to the front tarsi cover the eyes of the female.
Wittmann & Blochtein (1995) examined the fine detail of the
surface in the tarsal groove, where they found pores leading
to cuticular ducts, and evidence of pheromonal products,
mainly of carbohydrates and esters. Mid tibiae and hind
tarsi also have similar pores, suggesting the possibility of
chemical communication either in territorial marking or
in post-copulatory behaviour. The significant anatomical
adaptations in the male to restrict the movement of the
female are suggestive of conflicts of reproductive interest
between them, with coercive copulation on the part of the
male. However, Wittmann & Blochtein note instances of
females throwing males off by means of rapid abdominal
jerks. It would be interesting to compare the mating system
of species such as *M. willughbiella* with other species of leaf-
cutter, such as *Megachile leachella*, in which the males have
no such anatomical adaptations.

Summary

Accounts of the mate-searching and mating behaviours
of a small sample of the bees that occur in the British Isles
reveal a surprising degree of diversity. Males sometimes
remain at or return to nesting sites and attempt to intercept
and mate with freshly emerging females. In some mining

bees, the males may even dig into the ground to reach virgin females. Where numerous nests are aggregated in a confined area, there is intense competition among males, often resulting in a scramble as several males grapple with each other to gain a foot-hold on a single female. In other cases a hierarchy of dominance gives priority to one, usually larger, male among the competitors. Alternative male strategies involve attendance at foraging sites favoured by females (resource-based polygyny). In a few cases, males monopolise mating opportunities by aggressive territoriality, but within species this is highly variable, depending on environmental conditions, relative sizes and ages of competing males and their density at a site. Alternative strategies include regular patrolling with opportunistic attempts at copulation and adopting vantage points from which to pounce on females. In some other species (particularly in the genus *Andrena*) males scent-mark trails that are attractive to both males and females of the same species (non-resource based polygyny). Given common (but not fully tested) assumptions that females in most species mate only once, while males may be polygynous, theoretical expectations are that males should seek out receptive females at their natal nest sites in species that nest in large aggregations. Where nest sites are more dispersed, males might be expected to search for females at likely flowers, especially in oligolectic species. *Andrena florea* and *Macropis europaea* are British species that fit this model. In polylectic species, non-resource based patrolling might be expected to evolve. Territorial male behaviour might develop in each of these scenarios, depending, among other things, on the density of males and associated intensity of competition. However, these patterns have numerous exceptions, and in some species examination of female spermathecae has shown the prevalence of mating even before females have emerged from their natal nests (Paxton, 2005). Overall, it seems likely that pheromonal communication plays a large part in mate-searching, courtship and copulation in all species, but much more remains to be discovered.

oligolectic
collecting pollen from only a narrow range of plant species

polylectic
collecting pollen from a wide range of plant species. See section 5.4 (p. 111)

spermatheca (plural spermathecae)
the sac in the female abdomen in which sperm are stored after mating

3.3 Making and priming a nest

Bees, together with some groups of wasps and ants, are distinctive among the insects in making provision for their offspring. In the more elaborately social species, such as honeybees and bumblebees, a distinct worker caste nurtures the larvae of the dominant female. As we shall see, among the so-called 'solitary' bees there are some species that

Fig. 3.22 Root-plate of a fallen tree.

Fig. 3.23 Vertical exposure made by passing vehicles with six species of nesting bees.

Fig. 3.24 Edge of an excavation in a sand quarry, with nests of *Anthophora bimaculata*.

Fig. 3.25 Nesting habitat of *Andrena fulva* and *Andrena cineraria* in an East London park.

Fig. 3.26 Small mounds formed by a nesting aggregation of *Lasioglossum malachurum*.

show varying degrees of social cooperation. However, the most widespread pattern is for each female to find and, if necessary, prepare a nest site, and construct a series of brood cells, priming each one with provisions. Into each cell is laid an egg, the cell is sealed and another constructed. In almost all species there is no parental involvement in the subsequent development of the offspring. Although the nests of solitary bees contain much less food than the communal nests of social species, they still represent a significant prize for intruders or usurpers. In addition to the predations of many other types of animal, several groups of solitary bees themselves have evolved as dependants of those that laboriously make nests and store up food for their future offspring. These bees are sometimes referred to as cuckoos, but are more accurately termed cleptoparasites. These bees are discussed in Chapter 4, but here we focus on the species that build and stock their own nests.

Perhaps the majority of bees that occur in the British Isles establish their nests in burrows in the ground. Some show a preference for flat ground, others nest on slopes, especially south-facing ones, while others burrow into vertical exposures, such as cliffs, the root-plates of fallen trees, old quarry sites or the sides of deep grooves made by vehicle tracks in soft ground. Some species require loose, sandy soil for their excavations, while others prefer clay, and still others burrow into compacted ground alongside paths (Figs. 3.22–3.25). Often sites with ground devoid of vegetation, such as sand dunes or recently excavated aggregate quarries are preferred, but some species, such as the striking tawny mining bee (*Andrena fulva*), or the black-and-grey patterned *Andrena cineraria*, often nest in lawns. Especially where bees nest on level ground their excavations leave tell-tale pyramidal mounds of excavated sand or soil, but in other cases their burrows are dug from within cracks in dry ground and much less evident except when a female arrives or leaves (Fig. 3.26). The females of one species, the hairy-legged mining bee, *Dasypoda hirtipes*, have a long drape of hair on their hind legs which they use to brush away excavated sand as they dig (Fig. 3.27). Where a more-or-less extensive area of suitable ground is available, mining bees can form dense aggregations, with hundreds or even thousands of burrows in close proximity to one another. These aggregations may simply be the result of the suitability of the site, especially if the right conditions are quite scarce in an area. However, there is evidence that at least in some species female bees are drawn to nest

Fig. 3.27 A female *Dasypoda hirtipes* brushes away sand as she digs her nest burrow.

Fig. 3.28 A fence post used as nesting habitat by *Heriades truncorum*.

Fig. 3.29 A simple bee hotel.

in close proximity to others quite independently of this, using scent cues. This frequently results in aggregations occupying only a restricted part of an apparently suitable larger area (Butler, 1965). However they are formed, aggregations provide conditions under which various types and degrees of social cooperation might have evolved.

Other species adopt, and sometimes modify, existing cavities, such as hollow plant stems, in which to construct their brood cells. Yet others, such as the small black *Heriades truncorum*, occupy hollows left by wood-boring beetles in dead trees or fence-posts (Fig. 3.28). Cracks and holes in the mortar of old walls, chimney stacks and the like may also be adopted, especially by species such as *Anthophora plumipes* that are frequent residents in urban gardens. Provision of artificial nest sites, or bee hotels, with cut canes, tubes, blocks of wood with holes of various sizes drilled into them and so on are now increasingly popular in gardens, and they are usually very successful in attracting occupants (unlike commercially made bumblebee nests) (Fig. 3.29) . Three species of *Osmia* nest in snail shells, making their cells in a series around the spiral cavity inside. Another rare species, *Hylaeus pectoralis*, has the unique strategy of occupying the disused, cigar-shaped galls on *Phragmites* reeds left by the fly *Lipara lucens* (Fig. 3.30).

Burrows

Mining bees belong to several different groups, such as the numerous species belonging to the genus *Andrena*, the small, inconspicuous species of *Lasioglossum*, the plasterer bees of the genus *Colletes*, the flower bees (*Anthophora*), some of the leaf-cutters (*Megachile*), *Macropis europaea* and the distinctive, hairy-legged *Dasypoda hirtipes*. Females of the mining bees use their strong mandibles to dig out their burrows, funnelling out the excavated sand or soil with their legs. In most species there is a single burrow with side-branches, each of which leads to a single cell. An exception is *Macropis europaea*, in which the cells are clustered together. In some species each branch in turn is filled with excavated material when the cell has been closed, so that there is no need to carry surplus material back to the nest entrance.

If the nest, with its brood cells, is to house the successive life-history stages of its occupants for up to a year or even more, then it needs to be resilient in the face of a number of threats. These include the possibility of invasion by parasites or cleptoparasites, but also risks such as flooding or desiccation, and infection by fungi and bacteria. In many species, efficient foraging and priming of each cell, so reducing the time during which it remains open, is the main defence against attacks by parasites and cleptoparasites (Chapter 4). To prepare the cell for the other challenges, the females provide a waterproof lining that is also impermeable to fungi and bacteria. First, the soil or sand that forms the cell wall is tamped smooth and firm and then the female secretes a fluid from the Dufour's gland in her abdomen. This is applied to the cell wall and solidifies to form a protective coating. In *Colletes*, the material used is composed mainly of macrocyclic lactones, and is applied to the cell walls by the short, blunt, bilobed tongue that is distinctive of this group of bees (Fig. 3.31). After it has been applied, the material polymerises to form a tough, transparent cellulose-like film. A testimony to the effectiveness of the waterproofing properties of this material is provided by a large aggregation of *Colletes halophilus* at a sand-dune nature reserve on the Essex coast. High tides and storms during the autumn and winter of 2013/ 2014 resulted in complete inundation of the site by sea water. In late summer of 2014 the bees emerged in their usual numbers, apparently unaffected (B. Seago and own observations) (Fig. 3.32).

Other mining bees, such as *Andrena* and *Anthophora* species, also line their nests with chemical products of the Dufour's gland, but these are often quite different compounds (e.g. triglycerides in *Anthophora*) that form smooth, waxy materials that are integrated with the surrounding sand or soil. These species may be long- or

Fig. 3.30 A cigar gall induced by *Lipara lucens*, later to be occupied by *Hylaeus pectoralis*.

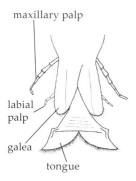

maxillary palp

labial palp

galea

tongue

Fig. 3.31 The bilobed tongue of a *Colletes* bee.

Fig. 3.32 Habitat of *Colletes halophilus* on the Essex coast.

short-tongued, but all have pointed tongues that could not be used to apply the cell lining. Instead, there is a raised and flattened structure on the dorsal surface of the final abdominal segment called the pygidium (or pygidial plate) which is used as a trowel for this purpose. Some species, notably *Dasypoda hirtipes*, do not line their brood cells. In the case of *Dasypoda*, the food mass provided for the larva is mounted on a stem from the cell floor, to minimise its contact with the cell walls. *Macropis europaea* females also do not secrete a material to line their cells, but instead line them with plant oils collected from yellow loosestrife.

Cavities

Females of other bee genera, mostly belonging to the family Megachilidae, such as *Anthidium, Osmia, Chelostoma, Heriades, Hoplitis* and some *Megachile*, make their nests in cavities of various kinds. The carpenter bees are also cavity nesters. There is only one established species of carpenter bee in the British Isles (*Ceratina cyanea*), but there are increasingly frequent reports of the very large, purple-black *Xylocopa violacea* and it may soon take up residence here. This species fully justifies the common name of carpenter bee, as it bites off wood fibres to make the partitions in its nest.

In common with many of the cavity-nesters, *Anthidium manicatum*, the wool-carder bee, is opportunistic in the nest sites it uses. These include hollow plant stems, exit-holes made by larger beetles and other insects in dead wood, crevices in old walls, window frames, and even metal chair arms (Severinghaus *et al.,* 1981). In general they make use of pre-existing cavities, which they line and seal with finely-shaved plant hairs. These are collected by the females from a range of plant species, including wild ones, such as yarrow, woundworts and mulleins. However, hirsute garden varieties, such as lamb's (or hare's) ears, *Stachys byzantina* and some cultivars of *Senecio*, are especially favoured. On *Stachys* a foraging female collects hairs from the densely clothed underside of a leaf, where she is invisible from above. She works astonishingly quickly on this plant, taking only twenty to thirty seconds to clip off a large mass of 'wool', held in a cage formed by the underside of her abdomen, legs and mandibles. Having collected a full load she leans back from the leaf, remaining attached by her hind legs and uses her mandibles, fore and mid legs to compress and shape the mass into a near-perfect sphere. She then flies from the leaf, but performs a brief orientation flight above the plants before flying directly off, presumably

Fig. 3.33 Patch of lamb's ears in an urban park with stages in the formation of a ball of 'wool' by a female *Anthidium manicatum*.

to a nest in process of construction. She returns in 5 to 10 minutes to repeat the operation (Fig. 3.33).

The leaf-cutters, *Megachile* species, have a range of nesting sites, including cavities in dead wood, plant stems and crevices in old walls as well as burrows in soil. The females line their cells with symmetrically shaped sections of leaves. These are cut out by a scissor-like action of the broad, sharp-edged mandibles from a range of plants, though rose leaves, and sometimes petals, seem to be the favourites (Fig. 3.34). The bee cradles the cut leaf-section between its legs and the ventral surface of its body as it flies back to its nest. Each cell is lined with ten or more pieces of leaf, forming a cylindrical container, and the sections are bound together with a secretion from the bee's salivary gland. The completed cell is finally sealed with an almost perfectly circular section of leaf.

The mason bees, *Osmia* species, utilise some of the most unusual nest sites. *Osmia bicornis* (formerly *Osmia rufa*) is perhaps the most familiar of all the solitary bees, as it readily takes to artificial nest sites in parks and gardens, as well as making its cells in crumbling mortar in old walls

Fig. 3.34 Symmetrical holes made by a leaf-cutter bee.

or chimney-stacks, or burrowing into soft ground or dead wood. Though it frequently uses ready-made cavities, it can excavate burrows in friable materials. The cells form a series along the cavity or burrow, the dividing walls between them being made of mud, and the nest is then also plugged with mud. This is collected by the female and carried back to the nest in the mandibles, and tamped into place with the two facial projections referred to in its current scientific name (*bicornis* = two horns) (Fig. 2.55). Some other common *Osmia* species (such as *Osmia leaiana* and *Osmia caerulescens*) have similar nesting habits, but instead of mud they use a mastic produced by chewing up plant fragments as cell dividers and plugs (Fig. 3.1).

Perhaps the most surprising options for nest sites are snail shells, the preferred solutions for three more of our *Osmia* species. The largest of these is *Osmia bicolor*, and its nesting habits have long been a topic of interest for ento-mologists. The females establish their nests in empty snail shells, such as those of the garden, banded and Roman snails (*Cornu aspersum, Cepaea nemoralis,* and *Helix pomatia*). Depending on the size of the shell, the bees make four or five cells within the spiral, each cell divided from its neighbour by a wall of chewed plant fragments. When the nest is complete, a band of fragments of earth, tiny stones, and shells separates the cells from a plug of chewed plant material at the mouth of the shell. The bee is common along rides in Kings Forest, in the Suffolk Brecks, where a local naturalist noticed a bee flying with a load 'like a witch on a broomstick', though the 'broomstick' would have been as long as a telegraph pole in proportion. Noticing the bee landing repeatedly with its loads at the same spot, he went

Fig. 3.35 A female *Osmia bicolor* covering its snail-shell nest with vegetation (R. Parker).

to investigate. A trellis was being constructed over a snail shell, consisting of last year's pine needles and strips of grass. Eventually 'the construction grew from a trellis to a wigwam, and the snail shell disappeared from view' (Parker, 2008) (Fig. 3.35). The great Victorian bee expert V. R. Perkins described this behaviour as long ago as 1884 (Perkins, 1884), and the website of the Bees, Wasps and Ants Recording Society has an outstanding series of video clips by John Walters, showing a male bee checking a snail shell, a female moving one around into the 'right' position, and then bringing in small twigs, bramble clippings and grasses to form a hood over the shell (bwars.com). Baldock (2008) reports as many as 50 females in an area of 10 m^2 nesting in snail shells and using beech bud scales as well as grass stems to cover them.

It is unclear whether the elaborate covers made for their nests serve to protect them from invasion by predators or parasites, or whether their main function is to prevent the nests from over-heating when exposed to the sun. Since the snails whose shells are used are of species favoured as food by thrushes, it would be interesting to know if bird predation would otherwise be a risk for *Osmia bicolor*. The other two *Osmia* species (*Osmia spinulosa* and *Osmia aurulenta*) that use snail shells do not seem to conceal them in the same way, although there is some evidence they may camouflage them with chewed leaf mastic (personal communication, N. Owens).

The females of the small metallic green-blue carpenter bee, *Ceratina cyanea*, make their nests in cut or broken stems of bramble or rose. These are usually horizontal, and often close to the ground. Cut stems placed on turf can be used to attract nesting females of this species in the areas of south-east England where they occur. The females use their mandibles to excavate a length of stem by cutting a burrow through the pith. The cells are 7–9 mm in length and 3 mm wide, and they are separated by walls of chewed pith fragments. Hollowed out stems are also used by the adults of both sexes as places to hibernate and collecting stems with small entrance holes cut into the pith is said to be the best way of finding the bee (Else, 1995).

Stocking the nest

Despite the great variety of nest sites and construction techniques, the bees have much of the rest of their nesting biology in common. As each cell is constructed, the female makes repeated trips to forage for pollen and nectar, returning to prime the cell with food for the future larva. There are differences between bee species in the consistency and in the constituents of the food mass. In species belonging to the families Halictidae, Andrenidae and Melittidae the mixture of pollen and nectar is firm, and moulded into a roughly spherical shape. Nectar is carried internally, but some of it is also used to compact the pollen as it is carried back to the nest. In *Anthophora* and *Megachile* species the food mass is viscous, and takes the shape of the part of the cell against which it is placed. In *Hylaeus* and *Colletes*, the primary ingredient is nectar, with pollen added, and so the food mass is liquid. The females of both genera line their cells with a transparent and waterproof film. In both *Dasypoda* and *Macropis* the food mass is mounted on a basal projection (Michener, 2007), and in *Ceratina* the food-mass is loaf-shaped (Else, 1995). In some species, most notably *Macropis europaea*, plant oils form part of the food mixture, as well as being used in the construction of the cells. Fatty compounds secreted by the Dufour's gland are added in some *Anthophora* species, but it is unclear whether this is a nutritional addition, or whether it has some pre-servative function (or both) (Batra & Norden, 1996). One egg is laid in each cell, either attached to the food mass, or to a cell wall.

In *Anthophora plumipes,* females alternate between two behavioural phases: searching and provisioning. They may take one or two days to locate a suitable nest site (the searching phase) and then turn to provisioning from five to eight cells. Stone *et al.* (1995) note the high demands on foraging females, carrying loads of from one third to a half of their body weight, and they estimate that a female might provision at most one cell in a day. Their reproductive success is highly dependent on the number of cells they can provision, and accordingly they forage even in poor weather, and at temperatures close to 0°C. Sexual harassment by males, whose reproductive interests are served by maximising the number of copulations they achieve, can significantly disrupt female foraging.

Similar behavioural patterns may apply in other species, too, although there is great variation among species in the quantities of material that can be transported in a single trip, the number of trips required to provision each cell, and the speed at which each trip can be completed. A review by Neff (2008) of reports on the provisioning behaviour of 92 species of solitary bee concluded that females of the great majority of them manage to provision no more than one cell per day, and sometimes less. The ability to provision more than three is rare. Apart from inclement weather, limiting factors include the availability of floral resources. Two studies of captive populations (of *Megachile* and *Osmia* species) indicated that other things being equal, increasing resource availability did increase the rate of provisioning and completing brood cells (Kim, 1999; Goodell, 2003). Also the transport capacity of the bee is a relevant factor. Measures of the weight of pollen collected by females of different species in a single foraging trip show that this varies greatly as a percentage of the body weight of the bee. In some *Megachile* species, for example, this is low compared with other bee genera, resulting in the need for more foraging trips to complete provisioning of a cell. This may be explained in terms of evolutionary selection of smaller *Megachile* females who can use more abundant small holes and crevices as nest sites. In addition, females of the nest-making Megachilidae have to provision their cells not just for the development of their larvae, but also for construction of the cocoons they spin. This is surprisingly costly, accounting for up to 50% of adult dry body weight, and implying that a bee belonging to these genera may have to collect two to three times its own fresh body weight to provision a female cell (Bosch & Vicens, 2002; Neff, 2008).

Less well-recognised is the role of the time taken for eggs to mature within the female, and consequently, the minimum interval between successive eggs being laid by a female. There is evidence that at least for some species this is a limiting factor on the rate at which cells are completed. Many studies of this are short-term, and, Neff (2008) argues, likely to greatly over-estimate the number of cells that can be completed over the active life of the female bee. In addition to foraging trips for pollen, nectar and/or plant oils to provision the nests, females spend time constructing cells, clearing out earth (in the case of mining bees), sealing cells and so on. Also, in many *Andrena* species periods spent foraging for pollen are followed by one or more days of foraging for nectar only, when, presumably, the bees are not actively constructing or provisioning cells (Neff & Simpson, 1997). It could be that there is a trade-off between egg-producing capacity and the devotion of metabolic effort to cell construction and provisioning, as most cleptoparasites have much higher rates of egg production.

Another limiting factor is the distance between the nest and a suitable foraging site. The foraging ranges of solitary bees are generally thought to be smaller than those of honeybees and bumblebees (Gathmann & Tscharntke, 2002). It should be noted that for so long as a cell remains open it is vulnerable to entry by a parasitic or cleptoparasitic enemy, so foraging efficiency is at a premium (see Chapter 4). As soon as provisioning is complete, the female seals the cell. In *Colletes* species, the sex pheromone linalool, which females secrete from the Dufour's gland, is also used as an anti-fungal and anti-bacterial agent, applied around the cell entrance before it is closed (O'Toole, 2013).

In all bees, the mode of sex-determination is haplodiploid. That is, females result from fertilised eggs, males from unfertilised ones. Females store sperm in a vessel called the spermatheca, which opens on to the ovipositor. So, as each egg is laid, they can determine the sex of the offspring by releasing or not releasing sperm. In most bees, males are significantly smaller than females, and emerge first. One implication is that those bees that build their cells in tubular-shaped hollows or tunnels have to place unfertilised eggs in cells closest to the entrance – or, from the point of view of the emerging bee, the exit. Another implication has to do with the quantity and quality of the food provided for the developing larva. As future male bees will generally be smaller than females, they require less provisions, although there is puzzling evidence that males are more expensive than females, per unit mass (Danforth, 1990). Since the female is under time-pressure to complete provisioning, and also has control over how much food is provided, she can, and would be expected to, adjust her foraging effort according to the sex of the eggs she lays in each cell.

There is evidence across a wide range of nest-building Hymenoptera that, among insects of the same species, larger females forage more efficiently, live longer and produce more eggs than smaller ones. There is also evidence (Siedelmann, 2006) that foraging efficiency in bees declines with age. In *Osmia bicornis,* females are on average some 1.6 times larger than males. Since the food store in each cell must provide all the nutrients required by the developing larva (including its cocoon, in those species that have them), and there is a direct relationship between nutrition and the size of the resulting adult, cells destined to yield male bees require less provisions than those destined to produce future females. Siedelmann *et al.* (2010) investigated the relationships between the size and age of females of a German population of *O. bicornis* and their provisioning of cells destined for male and female offspring, under risk of parasitism. Given these assumptions and the evidence of high rates of parasitic attack when the cells are open (over 31% of their large sample of provisioned cells were parasitised), the expectation was that larger females would bias their egg-laying in favour of daughters while smaller ones, at higher predation risk because of their slower foraging rate, would compensate by producing more sons. These expectations were confirmed, as also was the expectation that larger females would tend to produce a higher proportion of sons as they aged. In a few species, notably *Anthidium manicatum*, sexual selection for large size in males should lead to opposite sex-determination strategies on the part of egg-laying females: larger females should bias their production in favour of males, and smaller ones should favour daughters. The appropriate test remains to be carried out.

Navigating

As if selecting a nest site, and building and stocking the cells were not enough, the females face the challenges of finding and re-finding the resources they need for constructing and priming their cells, and relocating their nest entrances after each foraging trip. How do they do this? A clue is provided by the common observation of a bee leaving its nest (or a localised resource site) and performing a repeated zig-zag flight, with wider amplitude at each repetition before it flies up and away. The bees are now understood to process the visual information received during these orientation flights into cognitive maps on different spatial scales. As the bee returns it is able to direct its approach by recognising gross landscape features, and, as it gets closer, to locate its nest entrance in relation to nearby objects such as stones or plant-stems. Bees have yet another back-up means of navigation, especially useful for longer distance flights. This is orientation in relation to the position of the sun, which they can use in combination with their inbuilt clock, and which can be used even when the sun is obscured by cloud. O'Toole (2013) points out that the bee's ability to bring together all this information means it has, in its minute brain, the computing power of a modern laptop computer.

An accidental test of the resourcefulness of a female leaf-cutter in finding its nest entrance was provided by my own attempts to take photographs of one re-entering its nest with leaf cuttings. Several adjacent nests of two *Megachile* species (*Megachile leachella* and *Megachile willughbiella*) were in burrows along a low sandy bank by the coast. There were sporadic tufts of dry grasses and ribwort plantains.

The bees were leaving and returning at regular intervals, but my efforts at photography were frustrated by the speed at which they flew directly into their nest entrances with their cargo. In hopes of slowing one down, I roughed some sand over the entrance to one of the *M. willughbiella* nests. The bee duly arrived with a large section of leaf and flew to within 3–5 cm of the (now invisible) entrance and flew off. It returned twice, and repeated the approach and retreat. Next, it returned having deposited its load, and re-excavated its entrance. I still had no photos of it arriving loaded. My next trial was to leave the entrance open, but to place an obstacle – in the shape of a plantain leaf - in front of it. This time the bee arrived, appeared confused and retreated twice before eventually finding its way into the nest. Presumably these bees had already memorised the locations of their nest entrances, as under normal circumstances they did not perform an orientation flight on leaving. However, any alteration of the surroundings of the nest while the bee was inside triggered an orientation flight when it left (Fig. 3.36).

Although visual cues seem to be the main means of navigation used by bees, scent is also important for some purposes. For those bees, especially mining bees, which form large aggregations, there is often a distinctive scent. This is especially marked in *Colletes* species (Fig. 3.37), and the multi-functional compound linalool is primarily responsible. O'Toole (2013) describes his success in attracting males of *C. cunicularius* by anointing his hair and beard with this substance. As both males and females are attracted by the scent, it seems likely that it plays a part also

Fig. 3.36 A female *Megachile leachella* takes a leaf cutting into its nest; a female *Megachile willughbiella* makes a normal approach to its nest; the bee approaches a leaf placed so as to obstruct access to its nest; and the bee navigates the obstruction.

Fig. 3.37 A male *Colletes halophilus* cleans its antenna, an important organ for detecting chemical signals.

Fig. 3.38 A female *Colletes cunicularius* returns to its nest with a pollen load.

in enabling females to locate the aggregation within which their nest is situated. However, there must be a particular challenge involved in locating a specific nest entrance in a large aggregation. In some species females 'antennate' the nest entrance before entering, suggesting that an individually specific scent-blend may be involved. However, returning females of *Anthophora bimaculata* at an aggregation in the bank of an old quarry approached to within a few centimetres and performed a tight zig-zag orientation flight, followed by hovering briefly facing the array of nest entrances before flying into what was presumably the right one (own observation). This procedure suggests the use of very fine-grained visual cues (Fig. 3.38).

3.4 Solitary or social?

Bees and wasps display a range of life-styles from solitary to fully social (eusocial), with some being able to shift between social and solitary modes of reproduction according to environmental conditions. In some cases, genetic differences between regional populations of a single species may influence which life-style is adopted. This has made bees an especially valuable group of animals for researchers into the evolution of sociality, both in the sense of explaining the transition from solitary to social modes of life, and in the sense of the further changes in forms of social cooperation once it has become established in a lineage. In the halictid bees (i.e. in the British Isles, mainly the mining bees of just two genera, *Lasioglossum* and *Halictus*) social modes of life have evolved independently several times, and also have been lost in some lineages.

Fig. 3.39 A female *Andrena scotica* (=*Andrena jacobi*).

The widespread habit, especially among mining bees, of forming large aggregations is believed to be a condition that may favour the evolution of social behaviour. Fully social behaviour is rather uncommon among bees other than honeybees, bumblebees and their relatives, the mainly tropical or subtropical stingless bees. But various forms and degrees of social cooperation can be found in several species of *Andrena*, *Lasioglossum* and *Halictus* that occur in the British Isles. The most elementary form of sociality occurs in some species of *Andrena* mining bee, and takes the form of two or more females sharing a common nest entrance and burrow, but each building, provisioning and laying her eggs in distinct cells in her own branches of the main burrow. This is termed a communal association, and, in *Andrena scotica*, may involve as many as several hundred females sharing a nest (Paxton *et al.*, 1996) (Fig. 3.39). This elementary form of social cooperation is usually facultative – that is, the behaviour is found in some aggregations and not in others and is presumably influenced by environmental conditions. However, some species of *Andrena* that occur in the British Isles (e.g. *Andrena bucephala* and the very rare *Andrena ferox*) are obligate communal nesters (that is, the behaviour is found in all aggregations). There are other, intermediate levels of social cooperation, such as several females inhabiting the same nest and provisioning cells indiscriminately, without any specialised division of labour. A more complex form of semisocial cooperation involves several females of the same generation cohabiting in a nest, but with a division between some that have enlarged ovaries and sperm in their spermathecae and others that have less developed ovaries and are unmated. The former group do most of the egg-laying, while the latter do most of the foraging and nest construction. Semisocial colonies of this sort occur among American halictid species (O'Toole, 2013) though so far as is known this is not a pattern represented among halictids that occur in the British Isles. Some carpenter bees cohabit as a mother and her daughters, with the mother feeding the developing larvae (progressive feeding rather than mass provisioning as in other solitary bees) and continuing to guard them until they reach adulthood.

Eusocial forms of cooperation are found principally among the bumblebees and honeybees. Eusocial species are ones in which a dominant female (queen) monopolises egg-laying, and produces daughters who carry out various functions such as maintaining the nest, care of the immature offspring of the queen, foraging and defence from predation. These workers are a distinct caste, smaller than the queen, and generally do not mate or lay eggs (except when the queen becomes moribund or dies). At a later stage in the colony cycle the queen shifts from producing workers to laying eggs which are destined to become potential queens (gynes) or, in the case of unfertilised eggs, to become males. In honeybees the colony is perennial, and reproduction occurs by way of fertile queens leaving with a swarm of workers to found a new colony. This is regarded as the most advanced form of social cooperation in bees, while bumblebees and some halictid bees are also eusocial, but 'primitively' so. In these bees the nest typically lasts for one year, with only mated females surviving through the winter. In spring the female constructs a nest, forages and provisions a series of cells. In this phase she behaves as a solitary bee, hence the designation of this form of social life as primitive (although there is no suggestion that they are on route to a more 'advanced' stage). The eggs she lays in this first batch emerge as workers, and increasingly the female (queen) stays in the nest and devotes herself mainly to egg production and exercising control over the division of labour among the workers. At a certain point, she switches to laying the eggs that are destined to become the males and future queens.

This form of social cooperation is found in several halictid bees, the most thoroughly studied of which is *Lasioglossum malachurum*. This bee nests in dense aggregations, often with small pyramids of excavated soil around each nest entrance. In spring each fertilised female digs a burrow, and constructs and stocks cells. She then remains in the nest and seals the nest entrance until the offspring

Fig. 3.40 A worker *Lasioglossum malachurum* returns to the nest with pollen loads, as another leaves.

resulting from these eggs emerge as adult workers. As with bumblebees, the workers are in general smaller than the founding female (who now functions as queen of the nest), but they are usually much fewer in number than in bumblebees – on average only four or five to a nest. Another difference is that adult female *L. malachurum* are not prede- termined as queens or workers – according to time of year, size or interaction with nest-mates they may take on either role (Fig. 3.40). Once the first brood of workers has emerged the next phase in colony development begins with them foraging to provision cells in which the queen lays a further series of eggs that will go on to emerge as adult males, and reproductive females (gynes) with developed ovaries. As described above, males scramble to mount females as they emerge from their nests. The males die off before the onset of winter, while the females survive to begin the cycle over again the following spring.

The generally accepted explanation of the 'altruism' of the worker caste in caring for the queen's offspring, while sacrificing their own direct reproductive interests, has to do with the very close genetic relationship between the workers and the queen (as well as one another). The genetic relationship is especially close when the queen has mated with only one male. However, detailed research using DNA analysis to assess degrees of kinship in *L. malachurum* nests reveals a much more complex picture. In one study carried out in Germany, at least six out of 18 nests that were excavated contained alien workers – that is to say, workers who were not daughters of the resident queen (Paxton *et al.*, 2002). The study did not rule out the possibility that some of these might be arrivals from other nests, but the most likely explanation was that the current queen was a usurper, and the alien workers were the offspring of the founding queen. In these cases of mixed heritage in a nest, the dominant female appeared to be capable of suppressing the reproduction of her own offspring, but either tolerated or was unable to suppress the production of gyne-destined eggs by alien workers. Independently, *L. malachurum* workers with developed ovaries and full spermathecae have been observed while foraging, and workers as well as queens are considered capable of laying male-destined eggs (as well as female ones). This study also established that, although most queens of this species laid eggs fertilised by one father, females are capable of repeated matings under laboratory conditions, and in nature may mate with up to three males. As mentioned above, females of *L. malachurum* have different pheromonal blends after mating which reduce their attractiveness to subsequent males, but as the females control fertilisation of the eggs, a male may still benefit reproductively from mating with a previously mated female (e.g. if egg-laying follows the rule of using the sperm from the most recent mating). It seems likely that the queen's differential treatment of her own and alien workers is mediated by pheromonal communication.

During the solitary phase when the foundress female is building cells and foraging to provision them, the nest is vulnerable to usurpation by other females of the same species, and it has been thought that the females' practice of sealing the nest entrance and remaining inside while the first brood develops has evolved as a strategy to reduce the risk of usurpation. However, Zobel and Paxton (2007) conducted a study which revealed low and declining attempts at nest usurpation during the first cell-provisioning phase, as well as aggressive and almost always successful defence of the nest by resident females when this did happen. There are also potential costs to usurpation in terms of the likelihood of internal conflict from unrelated offspring of the original queen. The authors suggest an alternative plausible explanation of the behaviour of the resident female in terms of the relatively higher risk of foraging outside, compared with staying at home. Where a queen is found to be coexisting with unrelated workers later in the colony cycle, this is likely to be the result of an opportunist take-over of the nest, following the natural mortality of the foundress.

Geography provides yet another complication in the story. *L. malachurum* is a common and very widely distributed bee, and is eusocial in its habits everywhere. However, in southern Europe the pattern of social life differs significantly from that in northern Europe. Richards *et al.* (2005) studied a population in southern Greece in which two or three broods of workers were produced in each nest before the switch to production of males and gynes. The average size of the worker cohort in these nests – averaging 30 to 35 at peak - was thus far higher than in the northern European population discussed above. Although there was a small minority of nests in which queens had mated with more than one male, and one or two in which two queens coexisted in the same nest, the overwhelming majority of nests had a sole, singly mated queen.

The researchers calculated that queens in the southern population raised far more offspring (by a factor of approximately 10) with the help of their workers than they would have done on the basis of their own expenditure of time and energy. However, it was probably in the reproductive interests of the workers to make their own nests. The suggestion here is that the eusocial condition is maintained by the females' ability to control the activity of the workers and, in particular, to prevent them from laying eggs. The puzzle in this case is that within a population, the queen's ability to dominate the nest falls as the number of workers grows, but the queens in the southern European population appeared to have more control over their workers, despite the far greater numbers of potential rebels. One plausible explanation is that selective pressure on dominant females in the southern population has favoured queens with greater effectiveness in securing subservience among other nest-mates. It would be interesting to know if there are identifiable and relevant genetic differences between the females belonging to these geographically disparate populations.

A further study, this time of a population at a geographically intermediate location in Austria, revealed more complexities (Soro *et al.,* 2010). In this population there were two broods of workers prior to the final brood of sexually reproductive males and females. All nests analysed were the product of a single foundress female, but some of these had produced daughters from more than one mating (up to three), as in the more northerly German population. A quarter of the nests had an alien worker, which was unrelated to the foundress queen, but which, in a few cases, could be linked to other nests in the sample. In this case, then, it seems that workers drifted from one nest to another. Further, in almost half of the sample of nests, the foundress queen was absent by the time the sexually reproductive brood emerged. However, unlike in the German case, there was no significant difference in ovarian development between the workers who were daughters of the foundress queen, and unrelated nest-mates.

These findings pose a number of puzzles. Theoretical assumptions about the evolution of sociality in hymenopterans have tended to emphasise closeness of genetic relationships between non-reproductive workers and the dominant female. On the contrary, these studies suggest that the presence of unrelated workers is a common feature of *L. malachurum* nests, and also that in some populations a minority of queens is multiply mated, leading to further genetic diversity among the workers in their nests. Since there is evidence that kin and nest identity can be recognised by pheromonal cues (Ayasse *et al.*, 2001), it is not clear how or why some workers drift between nests, or why they are accepted into their new homes. Finally, in the German and Austrian populations it seems that queens had little or no control over the reproductive behaviour of at least some of the workers in their nest. Theoretical models suppose that for eusociality to evolve queens should be monandrous unless workers have their ability to mate fully suppressed (e.g. Andersson, 1984; Boomsma, 2007).

monandrous

of females, or mating systems, in which females mate only once

Some other species in the Halictidae have been found to be eusocial in some areas but solitary elsewhere. Two species of *Lasioglossum* (*Lasioglossum calceatum* and *Lasioglossum albipes*) and one of the genus *Halictus* (*Halictus rubicundus*) that occur widely in the British Isles have this character (Fig. 3.41). In general it seems that in populations at high altitudes, or northerly latitudes, which encounter less favourable climates and/ or shorter seasons, the bees are solitary in habit. In warmer, more favourable environments with longer seasons, local populations are primitively eusocial. The nest is established by a fertilised, over-wintered female, whose first brood is made up mostly of workers. There follows a social phase, in which workers cooperate to provision cells that are destined to produce sexually reproductive gynes and males.

An interesting question posed by these species is whether there is an inherited basis to the different modes of life, or whether any population has the required plasticity to respond with either mode, depending on environmental conditions. Research by Soro *et al.* (2010) takes us some way towards answering this question – though it turns out to be more complicated than it seems. Earlier research had provided either direct or indirect evidence of a genetic basis for the differences of social or solitary

Fig. 3.41 Female *Lasioglossum calceatum*.

life cycles between different populations in *Lasioglossum albipes*, and in *Halictus rubicundus* in North America. Soro *et al.* (2010) analysed genetic variation among nine geographically distinct populations of *H. rubicundus* across the British Isles. Five of these were northerly and/or at high altitudes, and were almost wholly solitary in habit, while the other four were at more southerly localities and were eusocial in habit. In this study, the degree of genetic differentiation between populations correlated with distance apart, with the barrier provided by the Irish Sea as an additional factor. By contrast, there was no significant correlation between the degree of genetic differentiation and whether a population was solitary or social, and there was evidence of recent or ongoing gene flow between solitary and social populations. These results are strongly suggestive of an environmental effect in the British and Irish populations of this species, but the research methodology of this study could not definitively rule out a genetic component, and the results seemed starkly at variance with those of parallel work on the same species in America.

The same research team (Field *et al.*, 2010) carried out experimental translocation of over-wintered fertile females from solitary source populations to warmer, southerly social localities, and *vice versa*. Females sourced from a social locality all became solitary at their less favourable northerly destination, while almost half of those from the solitary source population adopted the social colony cycle at the new destination (and others adopted an intermediate behaviour). These results do confirm the view that the British and Irish populations are capable of adopting either solitary or social life-styles depending on environmental conditions. A plausible explanation of the contrast with the American populations is that the solitary and social populations there belong to two distinct and specialised lineages, while the British and Irish ones belong to a third, European lineage. This would have spread to the British Isles relatively recently, and retained its plasticity with respect to social or solitary life, either because there has been too little time for distinct adapted lineages to form, or because gene flow between populations has offset local selective pressures toward climatic adaptation. It seems likely that the bee first evolved the eusocial habit, and secondarily acquired the ability to adopt an alternative, solitary mode of life as an adaptation to harsher or more unpredictable conditions.

4 Cuckoos in the nest

4.1 Introduction

The evolution of the habit of constructing and provisioning brood cells must have improved the survival chances of bee larvae, but it also involves risks. As with honeybees and bumblebees, which store large quantities of food in their nests, even the relatively modest provisions in the solitary bee's cell constitute quite a prize for any interloper capable of accessing it. Mites often live in bee nests as scavengers, but their populations sometimes expand to the point of consuming the food store of the bee larva, resulting in its death. A few beetle species and flies, such as satellite flies (*Senotainia*) and *Miltogramma* species, also invade the cells of some bee species. On a Norfolk heath, a female *Leucophora* species was observed following a female *Andrena clarkella* back to its nest, and then waiting for half an hour for the bee to exit (personal communication, N. Owens). Another group of enemies are the bee-flies (especially several species of *Bombylius* and *Villa*), whose larvae enter the brood cells and eat the bee larvae. The widespread and often common *Bombylius major* is a familiar sight in spring in urban parks and gardens (Fig. 4.1). The females flick out their eggs individually in places where their host species make their nests. The resulting larva finds its way into host nests, enters a brood cell and remains inactive until the host larva is full-grown. It then attaches itself to the host larva, feeding by sucking out its body fluids. Several species of *Andrena*, including *Andrena fulva* and *Andrena haemorrhoa* have been reported as host species (Stubbs & Drake, 2014; Paxton & Pohl, 1999). Another bee-fly, *Villa*

Fig. 4.1 The common bee-fly *Bombylius major*.

Fig. 4.2 A coastal bee-fly, *Villa modesta*.

modesta, occurs on coastal dunes (Fig. 4.2). Although little is known about its hosts or its life-history, it has been reared from the snail-shell nest of an *Osmia* species, and it is believed the larvae are internal parasites in the larvae of their hosts (Stubbs & Drake, 2014). On the Essex coast they are commonly seen flicking their eggs on sandy ground where several solitary bee species, notably *Megachile leachella* and *Anthophora bimaculata*, make their burrows (personal observation).

Some species of solitary wasps capture and anaesthetise bees, then carry them back as a food store for their larvae. The best-known of these is the bee-wolf, *Philanthus triangulum*, which preys on honeybees, but the common digger-wasp, *Cerceris rybyensis*, preys on several species of solitary bees (Figs. 4.3, 4.4). Reported host species include several species of *Lasioglossum*, *Halictus rubicundus* and some common *Andrena* species (Baldock, 2010). Birds, too, frequently consume the contents of the cells of solitary bees – one of the hazards associated with artificial bee hotels that are becoming increasingly popular in suburban gardens.

Fig. 4.3 The bee-wolf, *Philanthus triangulum*, with a honeybee.

Fig. 4.4 The hunting wasp, *Cerceris rybyensis* takes a female *Lasioglossum malachurum* into its burrow.

4.2 The cuckoo bees

As if these alien nest invaders were not enough of a problem, bees of several genera have evolved as cuckoos, or, more correctly, as cleptoparasites in the nests of other bees. The most well-known of these are the six cuckoos associated with the social bumblebees. However, the solitary bees also include more than 50 species of nest parasites, belonging to five distinct genera, drawn from three families (Halictidae, Megachilidae, and Apidae). Unlike the larvae of the cuckoo bumblebees, the larvae of the nest parasites of the solitary bees are not fed directly by the host adults. They feed on the food store deposited by the host female, usually killing the host larva first. Most cleptoparasites are quite specialised, and adapted to invade the nests of just one or a small range of related host species. The genus *Sphecodes* has 17 species recorded in the British Isles (including one reported from the Channel Islands only). They are small, relatively hairless, black bees, usually with large areas of red on the abdomen. These invade the brood cells of various species belonging to two genera, *Halictus* and *Lasioglossum*, of the same family (Halictidae) as themselves. There are two cleptoparasitic genera in the family Megachilidae: *Coelioxys* and *Stelis*. Seven species of *Coelioxys* have been recorded from the British Isles (including two that currently are found only in the Channel Islands). In most species the females have pointed tips to the abdomen that are used to insert their eggs into the cells of their hosts, which are various species of *Megachile* and *Anthophora*. The genus *Stelis* has four species that occur in the British Isles. They are small to medium-sized, and are quite rare. They attack the brood cells of species in several genera in their own family,

Megachilidae. In addition to the bumblebee nest parasites, there are three cleptoparasitic genera of solitary bees in the family Apidae. These are *Nomada, Epeolus* and *Melecta*. The genus *Nomada* has 34 species (including five that have been reported from the Channel Islands only) that occur in the British Isles, most of which attack the nests of mining bees of the genus *Andrena*. The genus *Epeolus* has only two British species, both of which are cleptoparasites of species belonging to the genus *Colletes*. The genus *Melecta* once had two British species, *Melecta albifrons* and *Melecta luctuosa*. The latter species is now presumed extinct in the British Isles but the former parasitises the nests of the much more widespread (but still mainly southern) *Anthophora plumipes*. The associations of hosts and cleptoparasites are complex and still relatively little-understood so there is plenty of room here for amateur observers to gather new knowledge.

Although the cleptoparasitic bees belong to several distinct families, and presumably evolved their parasitic life-style separately, they do show convergent features. These can mostly be understood as shared adaptations to the demands of this mode of life. First, having no requirement to carry pollen back to a nest, they tend to be much less hairy than their non-parasitic relatives, and the females do not have any pollen-carrying apparatus. Also the pollen brushes on the mid tibiae and mid femora used by the females of other bees to transfer pollen are absent in the cleptoparasites. There are other distinctive features that seem to be related to conflict with hosts. These include strong cuticle, spines, plates and ridges, especially protecting the neck and waist, and spines on the tibiae. They also commonly have more powerful stings. Plates on the basitarsus, and the pygidium, which are used in nest-cell construction, are reduced in most cleptoparasites. Finally, there are modifications to the tip of the abdomen in those cleptoparasites (such as *Coelioxys* species) which pierce the brood cell of the host in order to lay their eggs. Also, since they do not need to build and stock a new cell between laying successive eggs, the female cleptoparasites tend to have more mature egg-cells in their ovaries. Unlike other solitary bees, they also may lay two or more eggs in each host cell. The resulting larvae kill their conspecific cell-mates as well as the host larva. Many, but by no means all, cleptoparasites are slender in shape and resemble solitary wasps. This is especially true of the black-and-yellow banded species of *Nomada*, and the red-and-black patterned species of *Sphecodes*. Although the cuckoo bees

pygidium
a raised and flattened structure on the dorsal surface of the final abdominal segment

do not forage to provision nests, they do visit flowers for nectar. However, the best places to find them, especially the ones that parasitise the nests of ground-nesting bees, are around the entrances to the nests of their hosts. Where there are large, dense aggregations of host nests, as in the case of some *Andrena* species, their cuckoos (usually *Nomada* species) may be seen in their hundreds.

4.3 The cuckoo lifestyles

Genus *Sphecodes*
In most species of this genus that occur in the British Isles, only the fertilised females hibernate through the winter, emerging in spring and seeking out the nests of their hosts. These are usually species of *Halictus* or *Lasioglossum,* although a few parasitise the nests of *Andrena* species. *Sphecodes pellucidus*, for example, is the principal nest parasite of *Andrena barbilabris*. Females of *Sphecodes* species make an opening in the closed cells of their hosts and first kill the host egg, before laying one or more of their own. Sick *et al.* (1994) report a *Sphecodes* female entering a cell of *Lasioglossum malachurum* head-first, presumably to open the cell and kill the host egg. It then backed out and entered again, this time rear-end first, presumably to lay its own egg. The bee remained in the nest until the next day, when it left, having sealed the nest entrance. More recent observations by Bogusch *et al.* (2007) revealed diverse behavioural interactions between host bees and *Sphecodes* females. In some cases the cleptoparasite entered the host nest while the female was absent, but in others the two encountered one another when the host returned, or was already present in the nest. A few cases were recorded in which the host and *Sphecodes* female arrived together. In many of these observations there was no evidence of antagonistic interactions between the two, but in some cases there was defensive behaviour on the part of the host and in a few cases there was actual fighting and use of stings. These conflicts commonly resulted in the death of one or other of the females.

Several of the host species used by the cleptoparasites in this study were eusocial, but the parasites usually only attempted to gain entry to nests early in the development of the colony when only the foundress female was present. However, Bogusch *et al.* (2007) did observe some cases of parasitism of later, more advanced colonies of social hosts. Given that at least some of these potential hosts have guard workers, it is surprising that the parasites are able to gain entry at later stages of colony development. However, as we saw in the discussion of sociality in *Lasioglossum malachurum* (Chapter 3), alien workers, unrelated to the foundress female, have often been found in nests. It seems that nest parasites, too, are able to move between host nests, even when the latter are in their social phase.

Many species of *Sphecodes* are capable of parasitising the nests of several host species. Since host specialisation could offer advantages in terms of efficiency in recognising and entering host nests and overcoming host defences, it might seem surprising (as in the case of flower-visitation: see Chapter 5) that host specialisation is relatively uncommon. Habermannová *et al.* (2013) show evidence of host flexibility and frequent host-switches in the evolution of *Sphecodes* species, while Bogusch *et al.* (2007) show that, despite the wide spectrum of hosts used by many *Sphecodes* species, in at least two species there is host specialisation at the level of individual females. In their study of brood parasitism at sites in central and eastern Europe, Bogusch *et al.* (2007) observed the behaviour of *Sphecodes monilicornis* and *Sphecodes ephippius* (both of which are widespread in the British Isles) (Figs. 4.5, 4.6). These species were already known to be generalised in their selection of host species, but this study discovered even more host species being infested, making totals of 14 for *S. monilicornis* and 17 for *S. ephippius*. However, at the level of individual bees, there was strong evidence that females of each species were constant to a particular host species. The nest aggregations under study included mixtures of potential host species, so constancy could not be explained in terms of proximity. In some cases female cleptoparasites flew to nest sites 5 - 30 m away in pursuit of nests of the same host species

Fig. 4.5 A female *Sphecodes ephippius* leaves a host nest burrow.

Fig. 4.6 A male *Sphecodes ephippius* mounts a foraging female. Note the 'knobbly' antennae of the male.

as their previous victim. The mechanism underlying host constancy remains unknown: it could be inherited, the result of imprinting to the scent of the host during larval development, or (as in flower constancy: Chapter 5) the result of specialised learning.

Genus *Coelioxys*

The species that occur in the British Isles invade the nests of various species of *Anthophora* and *Megachile*. The females have very distinctive conically shaped abdomens which (with the exception of one species that occurs in the Channel Islands) terminate in rigid dagger-like points (Fig. 4.7). The female uses this to pierce the cap of a completed brood cell of its host species. An egg is laid though the opening, with about one third projecting into the cell. The resulting larvae have strong cuticle on the surface of their heads and, in second and third instars, large, curved mandibles which are used to kill (and sometimes, reportedly, eat) the egg or larva of the host. The developed larva forms a cocoon, as with other species in the family Megachilidae, and over-winters as a prepupa. One of the most widespread species is *Coelioxys elongata* whose host, *Megachile willughbiella*, is also widespread and common in gardens.

Genus *Stelis*

The four species that occur in the British Isles are all uncommon, and rather little-studied. They attack the nests of members of several genera in the family Megachilidae: *Osmia, Anthidium, Hoplitis* and *Heriades*. The females lay their eggs in the open cells of their hosts, returning repeatedly to the same nest, as the host female builds more cells. The resulting larva has normal-sized but sharp mandibles which are used to kill the egg or small larva of the host.

Fig. 4.7 Female *Coelioxys*, showing the shape of the tip of the abdomen.

The most widespread, but still very local, species is *Stelis punctulatissima*. Its main host is *Anthidium manicatum*. The rare bee of south-eastern England, *Heriades truncorum*, is the host species of *Stelis breviuscula*, which was not discovered as a British species until 1984 (Else, 1998). Since the host species seems to be expanding its range, *S. breviuscula* may become more widespread (Else 1998; Baldock, 2008) and should be looked for wherever its host occurs. So far most sightings have been of specimens on fence posts, or at yellow Asteraceae flowers such as ragwort. Baldock (2008) reports observing a female of *Stelis ornulata* (the rare nest parasite of the uncommon *Hoplitis claviventris*) following a female of the host species as it flew around a root-plate, presumably searching for a potential nest site. The strategy of lying in wait at foraging sites and then following host females back to their nests may be a widespread method by which cuckoo bees locate host nests, especially when these are not located in large aggregations.

Genus *Nomada*

All of the species of this genus that occur in the British Isles are cleptoparasites, and most of them use one or more species of *Andrena* as hosts. There are reports of females of some species following host females back to their nests. Baldock (2008) reports this behaviour in *Nomada argentata*, in relation to its host, *Andrena marginata*, and in *Nomada baccata*, whose host is *Andrena argentata* (Figs. 4.8, 4.9). However, in the latter case, Baldock also reports patrolling behaviour of the female nest parasite in search of host nests. The females of several species have been shown to detect

Fig. 4.8 A scarce nest parasite, *Nomada argentata*.

Fig. 4.9 Its host, *Andrena marginata*.

suitable host nests by scent and also to gain access to nests of their *Andrena* hosts by mimicking host nest-scents. Unmated females do not produce these chemical scents, but they are secreted by the mandibular glands of the males, and sprayed onto the females during mating (Cane, 1983; Tengö & Bergström, 1977b; Ayasse *et al.*, 2001; O'Toole, 2013). The eggs are laid in open cells of the host species, embedded in the cell wall, with only the tip exposed. The process of embedding the egg is aided by adaptations on the ventral surface of the 6th abdominal segment. This is a variable bi-lobed or blunt-ended structure, usually equipped with tiny hooks and stiff bristles (Fig. 4.10). Sometimes two or more eggs (up to six) are laid in a cell. The resulting larvae are equipped with large, powerful mandibles that are used to kill other *Nomada* eggs or larvae until only one remains. The egg or larva of the host species is then killed. *Nomada ruficornis* is a widespread species that attacks the nests of the common *Andrena haemorrhoa*. Both species are on the wing in spring. *Nomada baccata* and *Nomada rufipes* are both associated with heather heaths, where they parasitise the nests of their heather-specialist hosts, *Andrena argentata* and *Andrena fuscipes*, respectively. *Nomada goodeniana* has several hosts, but the large and widespread *Andrena nigroaenea* appears to be the main one.

Fig. 4.10 Underside of the tip of the abdomen of *Nomada marshamella*, showing hooked structure used in cutting open host cell wall.

Genus *Epeolus*

There are only two species of this genus in the British Isles, and both are cleptoparasites in the nests of various species of *Colletes*. The females fly low over nesting aggregations of their hosts, alighting to crawl between and enter nest entrances. They lay their eggs in open cells of the host, between layers of the cellophane-like lining, with the anterior end projecting into the cell. The resulting larva

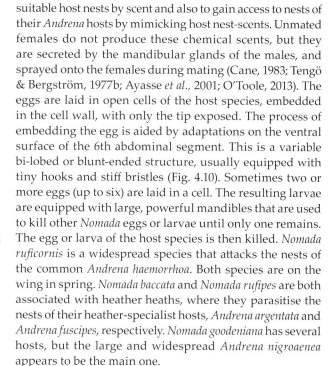

Fig. 4.11 *Nomada rufipes.*   **Fig. 4.12** *Andrena fuscipes*, its host.

has long, curved mandibles which it uses to kill the egg or young larva of its host. *Epeolus cruciger* is commonly associated with its host *Colletes succinctus* on heather heathland, while the very similar *Epeolus variegatus* takes several *Colletes* species, including *Colletes daviesanus*, *Colletes fodiens* and *Colletes halophilus*, as its hosts.

Genus *Melecta*

Melecta albifrons is a cleptoparasite of *Anthophora plumipes*. Its distribution mirrors that of its host – mainly southern and eastern counties of England, but widespread and not uncommon within that range. The female cuts a hole in the closed cell of its host, lays its egg, and then repairs the cell and re-plugs the cell entrance. Baldock (2008) reports observing five female *M. albifrons* mingling with females of *A. plumipes* at one of their nesting aggregations, and entering their nests. On two occasions he observed a female *A. plumipes* using her mandibles to drag a female of the cleptoparasite out of her nest. The first instar larva of *M. albifrons* has large, sickle-shaped mandibles and uses them to kill the host egg or larva. When full-grown it pupates in a cocoon. Males sometimes patrol in groups, and appear to scent-mark stopping points on their routes (Fig. 4.13). One was observed to dive at a foraging female, knocking it to the ground and then mating with it (see Chapter 3 for a more detailed description).

Fig. 4.13 A group of patrolling male *Melecta albifrons*.

5 Bees and flowers

5.1 Introduction

Most flowering plants – like most animals – reproduce sexually. Their sexual organs are contained in the flowers (Fig. 5.1). The female gametes, the ovules, are contained in one or more carpels, the male ones in pollen grains borne on anthers. Floral structures are immensely varied, but a simple and widespread pattern is a radially symmetrical array of brightly coloured petals, surrounding a central receptacle, which bears a number of carpels and from which arise a larger number of stamens. The carpels are flask-shaped and contain one or more ovules. At the apex of each carpel, usually carried on a narrow style, is a sticky platform, the stigma. Each stamen usually comprises a stalk, which carries a cigar-shaped anther within which large numbers of grains of pollen are formed. Below the petals, and often forming a protective cover for the bud in the early stages of development of the flower, is a radially symmetrical array of green sepals. Common deviations from this basic pattern are floral structures in which the petals (collectively known as the corolla) are fused together to form a tube and ones in which the shapes of the petals become differentiated from one another, so that radial symmetry is lost in favour of bilateral symmetry. For fertilisation to take place, pollen has to be transferred to the stigma, which is the receptive surface of the carpel. The flowers of most species, as just described, carry both male and female parts, but in most cases there are mechanisms that prevent or reduce the likelihood of self-fertilisation. The most obvious of these is timing: anthers release their

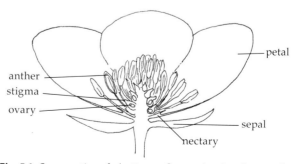

Fig. 5.1 Cross-section of a buttercup flower, showing the main floral structures.

pollen before the carpels are ready (or, sometimes, *vice versa*). In some species, pollen is incompatible with the ovules of the same plant, so the barrier to self-fertilisation is even stronger. Other species are able, in the absence of fertilisation by pollen from another plant, to be self-fertile. However, the result is usually poorer seed-set, qualitatively or quantitatively. In other plant species, each plant produces flowers of only one sex – white bryony (*Bryonia dioica*) is a familiar example.

Since plants are literally rooted to the spot, the result of these features of their sexual anatomy is that they depend on some independent agency to effect the transfer of pollen from one plant to another. For many plants (grasses for example) this agency is a feature of the physical environment, most commonly the wind, or, sometimes, water. But, for the great majority, pollen is transferred by animals. Estimates of the proportion of flowering plants dependent on animals for their reproduction on pollen transfer (pollination) by animals vary greatly. A recent study (Ollerton, Winfree & Tarrant, 2011), drawing on 42 surveys from across the world, estimated the overall percentage of animal-pollinated flowering plants to be 87.5%, but this varies according to latitude, with a slightly lower proportion (78%) in temperate-zone plant communities. The fact that so many flowering plants depend for their reproductive success on pollinating visits by animals clearly has great importance for the conservation of wild plants, as well as for the yield of many agricultural and horticultural crops (see Chapter 6).

Given their dependence on animal pollination, the evolution of flowering plants has gone hand-in-hand with that of their pollinators. Generally, this process of coevolution has resulted in mutual adaptations, such that pollinating animals are attracted and rewarded for their 'service'. Most commonly floral colours, shapes and scents are deployed to attract visitors, which, if they perform the appropriate movements, gain access to food in the form of nectar or pollen, or some other resource. The great variety of floral morphologies, colours, patterns and scents, together with the comparable diversity of apparent adaptations of flower-visitors for accessing the resources on offer, has suggested to observers that coevolution between plants and their pollinators has been at work in shaping close mutual associations between particular groups of plants and their preferred pollinators.

The concept of pollination syndromes has been widely

used to classify different floral patterns and link each with a group of pollinating visitors. So, for example, a hummingbird syndrome might be distinguished from a bee syndrome, or, more narrowly, a long-tongued bee from a short-tongued bee syndrome. Textbooks often feature quite spectacular reciprocal and exclusive adaptations between particular plants and pollinators, such as those between some orchids and their insect visitors.

However, as we shall see, more recent approaches emphasise that within the diffuse mutualism displayed by most pollination interactions, there are conflicts of interest between plant and pollinator, with associated asymmetries that extend in some cases to outright exploitation in either direction at the extremes. Flowering plants of many different families include species that attract pollinators but offer no reward. Orchids that lure male bees, flies or wasps to attempt copulation with them are the best-known examples. Among the orchids that occur in the British Isles, the early spider orchid, *Ophrys sphegodes*, relies on visits by a common solitary bee, *Andrena nigroaenea*, to trigger its pollination mechanism. The visual appearance of the lower petal (labellum) is said to mimic that of a female bee, but it has been shown that the surface also emits volatile chemical compounds (mixtures of alkanes, alkenes and other components), which mimic the pheromone bouquets of females of this *Andrena* species. It is this chemical mimicry that provokes the male bee into attempting copulation with the flower, and, in so doing, removing pollen (in pollinia) from an upper structure, the column. Pollen is transferred to another orchid if the same bee is induced into pseudocopulation again. Since the bees forage for food from a wide range of flowers, the reliance of the orchid on it as its sole pollinator seems very risky. In fact, only a small proportion of *Ophrys* plants are pollinated, but compensate by producing very large numbers of seeds (Scheistl *et al.,* 2000).

At the other extreme are insect visitors that directly consume nectar without brushing against the anthers. Bumblebees that bite holes in the base of the corolla tubes of deep flowers to take nectar, so by-passing the anthers, are a well-known example of 'nectar larceny' (Fig. 5.2). They have learned to get their reward without carrying out the pollination service that the plant requires. Other insects may continue to use the opening made by the bumblebee, so compounding its initial exploitation of the plant.

A vast range of animals visits flowers for the rewards they offer. Birds, most notably hummingbirds, and

mutualism
a pattern of interactions between individuals of two species from which both derive a benefit. Often species in mutualistic relationships do not benefit equally

pollinium (plural pollinia)
an adhesive mass of pollen grains suspended from the column of an orchid flower

Fig. 5.2 Nectar larceny. A bumblebee takes nectar from a hole in the side of a comfrey flower.

mammals, especially bats, are the main vertebrate visitors. However, in the British Isles and most of Europe, invertebrates, and especially some groups of insects, are overwhelmingly the main animal pollinators. Among the many groups of insects that gain some or all of their nutritional requirements from flowers are bush-crickets (Orthoptera), butterflies and moths (Lepidoptera), beetles (Coleoptera), flies (Diptera) of many distinct families, and Hymenoptera. Among the Hymenoptera, sawflies, wasps, ants and, of course, bees are all frequently observed flower visitors.

So, what do flowers have to offer? By far the most common reward is food, but there are exceptions. One, astonishingly complex, example is the pollination system of two families of Central and South American orchids. These produce scented oils that are collected and stored by finely adapted male orchid bees, or euglossines. Each male collects scents from a number of orchid species to produce a distinctive bouquet that it uses to attract females. The females, in turn, are important pollinators of forest trees (O'Toole, 2013).

The food resources provided by flowers are of three main kinds: pollen, nectar and, less commonly, plant oils. Pollen grains carry the male gametes of the plant, as well as nutrients whose function in plant reproduction is to power the journey within the carpels to fertilise the ovules. However, pollen collected by flower visitors is diverted and used as food for themselves or their offspring. This might seem to be a cost from the point of view of the plant, but, to the extent that plant reproduction is dependent on maintaining the services of pollinating visitors, there is an indirect benefit. Pollen is rich in proteins, but also contains smaller quantities of starch, sugars and important elements such as phosphorus.

While pollen collection involves removal of one of the reproductive elements of the plant, nectar is produced solely as a reward, and has no other part to play in the plant's reproduction. Nectar may be secreted by various floral components, but most commonly the nectaries are near the bases of the petals. However, where floral structures are complex, nectar may accumulate in pockets so placed as to require visitors to brush against the anthers in their efforts to reach it. Nectar consists of varying concentrations of sugars dissolved in water. Differing proportions of three sugars – sucrose, glucose and fructose – are found in the nectars produced by different plants, and these correlate loosely with their typical flower visitors: sucrose predominates in flowers visited by long-tongued bees or butterflies, and the other two sugars are most commonly associated with those visited by short-tongued bees and flies (Proctor, Yeo & Lack, 1996). But nectars also contain much smaller amounts of amino acids, which may be important to the nutrition of flower-visitors that do not consume pollen.

Some flowers offer plant oils instead of, or together with, nectar. Worldwide many groups of flowers provide oils as a reward for flower visitors, though this is rare among plants that occur in the British Isles. As we have seen, one very well-known example is yellow loosestrife (*Lysimachia vulgaris*) whose oils are collected along with pollen by the small black bee, *Macropis europaea,* and used to line its brood cells, as well as forming a part of the provision for its larvae.

Many insects, especially, visit flowers for nectar and/or pollen, but it does not follow that all of them contribute to pollination. As well as the bees that perpetrate nectar larceny (and other insects that make use of their work) there are more casual or, in some cases, relatively immobile species whose visits probably make little or no contribution to pollination. Some bush-crickets directly consume pollen, as do beetles of several groups, such as the tiny pollen beetles (family Nitidulidae), but in general these insects do not move from flower to flower frequently enough to be effective pollinators (Figs. 5.3, 5.4). Many adult insects visit flowers for nectar, primarily as a source of energy to sustain flight, or, to a limited extent, for body maintenance. These insects are the survivors of earlier stages in their life history, when, as larvae or nymphs, they took the nutrients required for their growth and development from non-floral sources such as plant tissue, dung, animal carcasses, or the internal

Fig. 5.3 A long-winged conehead bush cricket eats pollen.

Fig. 5.4 Pollen beetles share their feast with a bumblebee.

organs of hosts in the case of parasites. By contrast, bees have a more complex and intimate relationship to flowers, as the growth and developmental stages in their life histories are fuelled solely by the rewards offered by flowers: nectar, pollen, and, sometimes, plant oils. The provision of pollen in quantities sufficient to meet the full nutritional requirements of bee larvae, well beyond what might be necessary for the narrowly reproductive function of pollen, puts particularly strong resource demands on flowering plants, especially where these grow in low-nutrient substrates. That plants have evolved to provide these rewards suggests that the bees are more effective, or more reliable, pollinators than other groups of insects. This is confirmed by parallel aspects of bee biology. The branched (plumose) body-hairs that are characteristic of bees readily detach pollen and carry it from flower to flower. Many bees are capable of raising their body temperature well above ambient, and so can forage earlier and later in the day, and under more adverse weather conditions than most other insects. Also, the mere fact that they forage for pollen means that they engage physically with the parts of the flower much more actively than do many insects that simply forage for nectar, and only incidentally dislodge quantities of pollen onto their bodies. Very long-tongued insects such as butterflies and some hawk-moths, for example, may not even alight on the flowers they visit (Fig. 5.5).

But pollination requires more than simply carrying pollen from one flower to another. Cross-pollination can occur only if the pollinator carries pollen from flowers of one plant to those of another plant of the same species. This requirement suggests there might be strong co-evolutionary pressures towards specialisation, with a particular bee species becoming adapted to visit and access the resources of particular plant species, and plants developing floral structures that restrict access to just one, or a small number, of pollinator species that will have had little option but to have previously foraged from a plant of the same species. Reciprocally, the bee would gain a monopoly of the rewards provided by its favoured plant.

Fig. 5.5 A hummingbird hawk-moth nectaring at a distance.

5.2 Flower form and insect visitors

In fact, tight, reciprocal mutualisms of this sort are very rare, but in a more diffuse way, there do seem to be definite patterns of association between types of flower and particular groups of visitors, bees included. As mentioned above, different floral structures can be classified into so

Fig. 5.6 Harebell, a bell-shaped blossom, visited by *Megachile willughbiella*.

inflorescence
a group or cluster of flowers carried on a single stem, which may be branched as in Apiaceae (e.g. cow parsley, hogweed), or unbranched and bearing terminal compounds of many florets as in many Asteraceae (e.g. dandelion, ragwort, daisy)

Fig. 5.7 Valerian, a compound of tube-shaped florets, with *Anthophora quadrimaculata*.

many pollination syndromes, with the expectation that each is likely to be visited preferentially by a sub-set of pollinators whose size, shape, physiology, and behavioural abilities are adapted to perceiving and accessing the rewards on offer. The grouping of floral structures into pollination syndromes cuts across standard descent-based classifications into families, genera and species. So, flowers of plants belonging to several different families may be put into the same syndrome, on the basis of the similar functional organisation of their parts in relation to an expected group of pollinators.

Faegri and van der Pijl (1979) provided a classification of syndromes that has proved very influential and the following six types are of particular relevance to pollination by bee visitors. First is the most basic floral structure, as described above: radially symmetrical, with an array of brightly coloured petals surrounding a central receptacle bearing carpels and anthers, and with exposed nectaries, usually close to the centre of the flower. These are bowl or dish-shaped blossoms, and may be composed of single flowers, as in mallow, buttercups or bramble (Fig. 5.1), or may be formed by compound inflorescences, such as daisies. In the second group, bell- or funnel-shaped blossoms, the petals are fused together so as to enclose the anthers, carpels and nectaries. Typically, visitors have to enter the flower and brush against anthers and stigmas in order to reach the nectar reward. Bindweed (*Calystegia* species), most bellflowers (*Campanula* species) and bluebell (*Hyacinthoides non-scripta*) are examples (Fig. 5.6). A third group are tube-shaped blossoms. Like bells and funnels, they are radially symmetrical with fused petals, but the sexual parts and the nectar are enclosed in a narrower corolla tube that excludes entry by many foragers. Flowers of *Hebe, Buddleja* and other garden shrubs are assigned to this group, and they often form elements in more complex compound blossoms such as *Verbena*, teasels (*Dipsacus* species), scabious (*Knautia, Scabiosa* and *Succisa* species) and knapweeds (*Centaurea* species) (Fig. 5.7).

Two further floral syndromes that are more complex and are especially attractive to bees are gullet-type and flag-type. These are both bilaterally symmetrical (zygomorphic), and have independently modified petals that suggest adaptation to animal visitors to them. Gullet-type flowers include many members of the family Lamiaceae such as deadnettles (*Lamium* species), black horehound (*Ballota nigra*), sages (*Phlomis* species) and

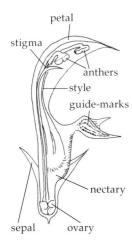

petal

stigma

anthers

style

guide-marks

nectary

sepal   ovary

Fig. 5.8 Cross-section of a flower of yellow archangel, a gullet-shaped blossom.

stigma

the receptive tip of a carpel where pollen germinates

Fig. 5.9 Flowers of ground ivy, a gullet-shaped blossom.

others, and the family Scrophulariaceae, including red bartsia (*Odontites vernus*), figworts (*Scrophularia* species), toadflax (*Linaria* species), snapdragon (*Antirrhinum majus*), and foxgloves (*Digitalis* species) (Figs. 5.8, 5.9). These usually have a tube- or flask-shaped base, which contains the nectar, and a lower petal expanded to form a platform on which an insect can settle. This is often distinctively coloured or patterned. The anthers and stigmas are positioned in the upper part of the fused corolla and brush against the insect visitor's back as it reaches into the flower for the nectar.

Flag-type blossoms are characteristic of members of the pea family (Fabaceae), such as trefoils (*Lotus* species), vetches (*Vicia* species), brooms (*Cytisus* species) and restharrows (*Ononis* species), and as elements in compound inflorescences such as clovers (*Trifolium* species). In Fabaceae an upper petal is expanded and brightly coloured (the flag). It is held to be the advertisement to potential visitors, while the lower petals, which form a boat-shaped enclosure (the keel) for the sexual parts of the flower, serve as a platform (Figs. 5.10, 5.11). When the flower is approached legitimately, the weight of the insect visitor on the keel exposes the anthers and stigmas, which then brush against the ventral surface of the insect's body. Flowers of broom represent a variation on this theme – they offer only pollen as a reward, and some of the exposed anthers curve up and over, so as to brush against the back of the insect.

A final syndrome is referred to as 'brush-type'. In flowers in this group the petals are usually greatly reduced or absent, and the sexual parts of the flower are exposed. The flowers are often borne in dense inflorescences and insect visitors simply scramble over them, seeking nectar or pollen. Important examples for bees and many other insect groups are the catkins of sallows or willows (*Salix* species) and ivy blossom (*Hedera helix*). Of course, this classification has obvious limits, some blossoms appearing intermediate between syndromes, and others consisting of compounds of two or more types.

The idea of pollination syndromes suggests that the various patterns of structure, colouration and scent, together with the floral rewards on offer, have evolved under selective pressure to attract frequent and effective visitors from the standpoint of their need for pollination. The floral traits of each type of flowering plant, on this set of assumptions, should be such as to attract the services of their 'most effective pollinator' (Stebbins, 1970). This leads

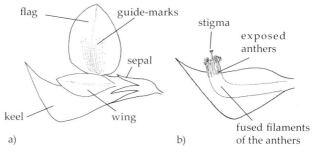

Fig. 5.10 (a) A side-view of a flower of spiny restharrow, a flag-type blossom and (b) a cross-section of the keel of a flower that has been visited by an insect.

Fig. 5.11 Flowers of spiny restharrow. Note the exposed anthers in the flower on the right.

guild

a group of species that use the same class of environmental resources in a similar way. They may not be otherwise closely related

to the expectation of some degree of association between groups of pollinators and particular syndromes, but it is not clear just how specialised these associations should be. Would selection have led to tight mutualisms at the level of species-to-species interactions, or to looser associations, such as between long-tongued bees and flowers with deep corolla tubes? Observation of actual pollinating activity reveals that most flowers are visited by a wide spectrum of insects, and most insect-visitors are able to access a wide range of floral syndromes. A further difficulty for the idea is that it is difficult to test, as many insect visitors do not contribute to pollination, and it is difficult to establish among even those that do, which one might be the most effective (In every season? At all points in the geographical range of each species? And so on.). On theoretical grounds it has also been argued that floral traits may be adapted to attract a less effective pollinator if that involves little or no loss of attractiveness to the most effective one (Aigner, 2006).

Corbet (2006) provides a three-fold classification of flowers and their visitors that retains something of the intuitive plausibility of the idea of pollination syndromes, while acknowledging the evidence of widespread generalisation. She draws on a very large data-set of observed flower visits (Ellis & Ellis-Adams, 1993) to advocate a division between allophilous, hemiphilous and euphilous plant floral structures. The first are shallow flowers, with relatively small nectar rewards, but open nectaries, while the second have semi-concealed nectaries with moderate supplies of nectar. The euphilous flowers are perhaps the most distinctive category, having a deep corolla and abundant, concealed nectar. Corresponding to these floral categories are three groups, or guilds, of flower-visitor. Developing the analysis of foraging behaviour given in

Corbet *et al.* (1995), Corbet (2006) distinguishes allotropous, hemitropous and eutropous flower visitors. The allotropous visitors are small, short-tongued insects with little ability to raise their temperature above ambient. They include many flies, beetles and short-tongued Hymenoptera. The hemitropous visitors include some larger insects such as hoverflies, bee-flies and butterflies, while the eutropous group includes larger insects such as long-tongued bees, butterflies and hawk-moths. However, as Corbet notes, even this more open system of classification has its limitations. Some deep euphilous flowers may have their anthers more exposed than their nectar, so allotropous insects may still visit for pollen even if they may not reach the nectar. Also, a sufficiently small insect may be able to access nectar stored in a deep corolla by entering with its whole body and there are some cornucopia flowers that attract an extra-wide spectrum of visitors with copious nectar which is readily accessible to both short- and long-tongued insects. These are frequently blossoms in which many florets are combined together in a single compound head, such as in Asteraceae, or a plate-like umbel as in Apiaceae, allowing the visitor to probe many florets without the energetic costs of flight. Examples are hogweed (*Heracleum*) with 436 visitor species (Fig. 5.12), dandelion (*Taraxacum*) with 375 and carrot (*Daucus*) with 314. Corbet cites a study of pollination webs in two meadows (Dicks *et al.*, 2002) as offering some empirical support for distinguishing at least two distinct compartments: bees and butterflies as visitors to euphilous flowers, and flies to hemiphilous ones.

Fig. 5.12 An inflorescence of hogweed: an insect cornucopia.

5.3 The bees

For a plant that depends on cross-pollination by insects, there must be a significant probability that an insect visitor previously visited a flower of a plant of the same species. For the reproduction of the plant to be successful, this requirement must be met in some way. In view of the difficulties faced by scientists in establishing clear links between specific flower types and specific groups of pollinators, there remain questions about how plants manage to reproduce successfully. These questions lead us to the behaviour of the pollinators themselves: in our case, the bees.

Costs and benefits

As we have seen, bees do not set out to pollinate flowers. For them, foraging is about obtaining nutrition, either for themselves or for their future larvae, or both. For many years research on bee foraging focussed on the energy costs of flight and flower-handling, in relation to the calorific value of the nectar collected or consumed. The classic work in this tradition was Bernd Heinrich's *Bumblebee Economics* (1979). His research centred on the foraging behaviour of bumblebees, and, as his title suggests, made use of an economic analogy with cost-benefit analysis. Neo-Darwinian evolutionary theory predicted that bees (like any other forager) should have evolved behavioural strategies to maximise their energetic gain per unit of energy expenditure. This set of expectations derived from optimal foraging theory has been tested by research into bee behaviour with mixed results. Bumblebee species that have been tested seem to fly much further between nest and foraging site than the theory implies, while patterns of foraging from complex inflorescences, and decisions when to switch from one floral resource to another, or one patch to another, do not all conform to theoretical expectations (see Benton, 2006, chapter 6).

Critics of optimal foraging theory have emphasised the need to build into the account of foraging behaviour other aspects of the ecology, morphology and behaviour of the group of animals concerned. There are three sorts of consideration that apply particularly to both social and solitary bees. One is that they are central place foragers. This means that they must fly between their nest and the patches of flowers from which they take resources. Various factors may limit their ability to minimise that distance. The most obvious one is that suitable nest sites may be scarce and not located with convenient access to a nearby place to feed.

Mining bees, for example, often need specific soil conditions, physical aspects and so on if they are to dig their burrows. Second, bees are especially vulnerable to parasites and parasitoids, as well as to cleptoparasites that invade their nest-cells. It would be surprising if their foraging strategies were not strongly constrained by avoidance of these risks (Dukas, 2001). For example, the need to evade trailing nest parasites may lead bees to avoid returning directly to their nests. Third, all bees must forage for nectar to fuel their flights, and, for many, to raise their body temperature above the ambient. However, for bees (except the cleptoparasites) nectar is not the only reward. Pollen is crucial to the diet of their larvae, and nectar is also an important source of water. Depending on the purpose of a particular foraging trip, a female bee may forage suboptimally for nectar because its main aim is to collect pollen to provision a nest cell, or water to achieve the right consistency of the food mass. There is a fourth consideration that applies to social species such as bumblebees, and also the small number of other eusocial species such as *Lasioglossum malachurum*, but not to the other truly solitary bees. This is that the reproductive unit is the social group of each nest, comprising a queen, workers, males and daughter queens. The foraging behaviour of each worker should be understood in terms of the overall needs of its nest-mates, not just its own nutritional requirements.

Various attempts have been made to quantify predictions from foraging theory using highly complex mathematical models that attempt to take these considerations into account. They go beyond what we need here, but it is still worthwhile noting the diversity of cross-cutting pressures that shape the foraging behaviour of bees as individuals or as species. As well as the various external conditions (including the availability of suitable flowers) that shape foraging behaviour, the full range of characteristics of the bees themselves – perceptual and learning abilities, innate preferences, cognition and memory, morphology and physiology, and life-cycle – may serve to enable or constrain foraging behaviour.

Finding the right flowers

Relatively little research has been done on the psychology of solitary bees, compared with the more popular honeybees and bumblebees (Chittka & Thomson, 2001). Such evidence as exists suggests broad similarities across all bee groups in their perceptual abilities and in the way these are used in foraging. The compound eyes of bees are effective at detecting movement and shapes. Their ability to discriminate between shades of colour, too, is acute. However, the colour spectrum to which they are sensitive differs from that of the human eye. The bee eye has cells sensitive to three colour groups – yellow, blue and ultra-violet. Bees reliably discriminate between these three basic colours, and, somewhat less reliably, can discriminate between shades within and between these, as well as combinations of them. The particular sensitivity of bees to ultra-violet light causes flowers to stand out against a less reflective background of green vegetation or soil, and it also means that colours and patterns of flowers can look very different to a bee, compared with a human view. Many flowers have visual guide-marks on their petals that help to guide visitors to nectar, and at the same time regulate their behaviour in relation to pollination. If the ultra-violet spectrum is included, many more of these appear. They are also more frequent in flowers that have a more complex structure, and which are therefore more demanding for insects to handle.

At a distance, bees are attracted to floral displays primarily by visual cues – colour and shape – but as they get closer scent plays an increasingly significant part. In experimental studies, bees that have been foraging on arrays of real or artificial flowers reject them if unfamiliar scents are added. The antennae are the main organs carrying scent-sensitive receptors, and at close quarters bees can detect and follow a scent gradient towards a flower, and locate the direction of a source of a scent by comparison of inputs from the two antennae (Raguso, 2001). Frequently scent-trails within the flower follow the lines of visual guides, complementing visual information. Workers of social bee species also leave scent cues on flowers to provide negative or positive signals to fellow workers, but solitary bees, too, deposit chemical signals on flowers. Females of *Macropis europaea* deposit scents that are attractive both to other females and to males, and it seems these relate to the distinctive mating system of this bee (see above, Chapter 3).

There is some evidence for innate preferences for specific colours and scents, but experimental work shows a high degree of learning and memory in the establishment of foraging routines (Dobson, 1987; Milet-Pinheiro *et al.*, 2013). Details of the location of the nest, the route to and from foraging patches, as well as which flowers contain rewards and the mechanics of accessing them must all be to a significant degree learned. Some species follow regular foraging routes on each trip, so presumably are able to steer according to directional cues, recall and recognise landmarks and so on. Honeybees are known to be able to steer by the position of the sun, adjusted by time of day, and there appears to be no reason why solitary bees should not have a comparable ability: the key tasks of regular return flights between forage patches and the nest are the same.

Accessing the reward and carrying it home

When the purpose of a flower-visit is collection of nectar, the labrum and mandibles open to release the complex tongue, or proboscis. This is equipped with taste receptors, and these, combined with tactile, visual and scent cues enable the bee to probe the nectar store of the flower, and draw up nectar through the tongue into the gut. Much of the discussion of mutual adaptations between flowers and pollinators refers to the significance of the tongue-length of flower-visitors. The distinction between long- and short-tongued bees is, of course, relative to the overall size of the bee. However, the variation between bees of different genera and species in the relative length of their tongues is very considerable (Figs. 5.13–5.20). *Anthophora* and *Megachile* species are often very long-tongued, and compare with bumblebees in this respect. They are able to access deep flowers such as many varieties of Lamiaceae (deadnettles) and Fabaceae (vetches, clovers and their relatives). Species of *Andrena* have relatively short, pointed tongues, though some species are large and so can still forage from some deep flowers. Bees in the family Colletidae (*Colletes* and *Hylaeus* in Britain) have very different, short, bi-lobed tongues (Fig. 3.31) that are used in cell construction as well as for collecting nectar.

Both male and female bees consume nectar to sustain

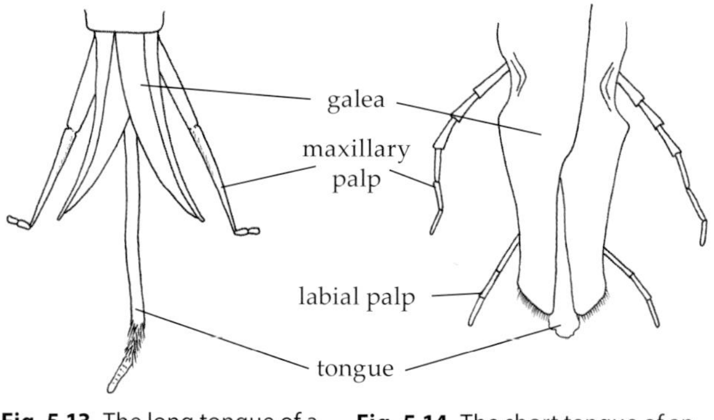

labrum
a variably shaped facial plate, situated below, and articulated with, the clypeus, and usually concealing the base of the tongue. See Figs. 2.14, 2.18 and 2.20

galea
maxillary palp
labial palp
tongue

Fig. 5.13 The long tongue of a *Megachile* species.

Fig. 5.14 The short tongue of an *Andrena* species.

Fig. 5.15 A male *Anthophora plumipes* uses its long tongue to probe the narrow tube of a green alkanet flower.

Fig. 5.16 *Anthophora plumipes* is able to legitimately access the long corolla of comfrey flowers.

Fig. 5.17 A female *Megachile leachella* uses its long tongue to access the nectar in a spiny restharrow flower.

Fig. 5.18 A male *Andrena* uses its short tongue to suck nectar from florets of alexanders.

Fig. 5.19 A female *Andrena fuscipes* forages from the shallow flowers of ling.

Fig. 5.20 A female *Andrena pilipes* forages from *Eryngium*.

their foraging activities, but for females who forage to provision brood cells, a surplus over these requirements has to be imbibed. In some species this nectar is used to moisten pollen loads, but more generally it is taken back to the nest and regurgitated to form part of the food store for the future larvae. Females of *Macropis europaea* use plant oils as an energy food for their larvae, and also as a component in the water-proof lining of their brood cells.

Fig. 5.21 A female *Macropis europaea* with a full pollen load.

Oils are collected from glands on the surface of yellow loosestrife petals by means of fine hairs on the underside of the smaller tarsal segments of the fore and mid legs, and used to consolidate pollen loads on the broadened hind tibiae and basal segments of the hind feet. These leg segments are densely clothed with long, backward-directed hairs which are loaded with the mixture of pollen and oils. However, the adhesion bestowed by the oils allows for enormous loads to be carried, far exceeding what could be collected in the pollen hairs alone (Fig. 5.21).

Plumose hairs clothing the body are a distinctive and general trait of bees, and are presumed to be an adaptation to pollen collection. This is consistent with the fact that these hairs are less dense in most species that have adopted a cleptoparasitic life-style, and so do not need to collect pollen. In some species that are covered by a dense coat of hair this also contributes to maintaining body temperature above ambient levels (see below). Some aspects of pollen collection are common across the different bee families and genera. The activity of the bee as it visits a flower, whether in search of nectar or pollen, typically causes it to brush against the anthers, and pollen is caught among the hairs on its body.

If the bee is on a pollen-foraging trip, its fore and mid legs are used to scrape the pollen off body hairs and onto some collecting apparatus so that it can be carried back to the nest. As is well-known, bumblebees and honeybees have a pollen-basket, or corbicula, on each hind tibia: this consists of a flattened and hairless outer surface fringed

Fig. 5.22 A female *Hylaeus* with a ball of nectar and pollen, about to be swallowed.

by long, stiff, inwardly curved hairs. Pollen is progressively loaded onto the corbiculae as the bee forages, and is taken back to the nest when it is full. The solitary bees do not possess corbiculae, but instead have a great range of alternative devices. Species in the genus *Hylaeus*, the yellow-faced bees, are distinctive in having no external adaptations for carrying pollen back to their nests, but carry it internally, already mixed with nectar in the gut (Fig. 5.22). Apart from these, and the cleptoparasites, females of other species divide into those that use hair-brushes on their hind legs (scopae) and those that collect pollen on the underside of the abdominal segments.

The former group comprises members of some twelve genera in the British Isles, including *Andrena, Colletes, Lasioglossum, Anthophora, Macropis, Dasypoda, Melitta, Eucera* and *Anthophora*. Those that carry pollen among brushes of long, stiff hairs on the underside of the abdomen belong to the family Megachilidae, and include members of the genera *Anthidium, Osmia, Chelostoma, Heriades* and *Megachile* among others.

Most species have brushes of stiff hairs on the inner surface of the enlarged first segment of the tarsi. These are used to remove pollen from body-hairs. Pollen may also be collected directly from anthers using the fore tarsi, and passed forwards to the mouthparts if it is to be directly consumed. If it is to be carried back to the nest, it is passed back via pollen brushes on the mid femora and tarsi to be collected on the hind legs or under the abdomen. Among those species that collect pollen on their hind legs, species of *Andrena* have some of the most complex adaptations. The outer surfaces of the hind tibiae and expanded first tarsal segments (basitarsi) are covered with long, backward-directed hairs, some of which are finely branched. Two species that collect the large pollen grains of scabious (*Andrena hattorfiana* and *Andrena marginata*) have feathery, as distinct from merely branched pollen-hairs (N. Owens, observation) (Fig. 5.23). The ventral surface of the femur is bare, but with fringes of long, inwardly curving hairs that form a pollen-basket. This is further enclosed by a loose tuft of very long, curved and feathery hairs (the floccus) that arises from the next leg-segment, the trochanter, and extends as a fine network over the femoral pollen basket. *Andrena* species also have an extra pollen-carrying device on each side of the propodeum. This is formed of a bare, or sparsely haired area of cuticle, fringed by long, curved hairs (Fig. 5.24).

propodeum
the modified first abdominal segment, fused with the thorax. See Chapter 1

The pollen-carrying adaptations of *Lasioglossum* and *Colletes* females are very similar to those of *Andrena*. In both genera the pollen is carried on the hind legs, which are equipped with arrangements of long, often feathery hairs on the tarsi, tibiae and femora (Fig. 5.25). However, in these groups, the floccus arises from the base of the femur, rather than from the trochanter, and in *Colletes,* pollen is often collected on both surfaces of the hind tibiae. In our sole species of *Dasypoda,* the hairy-legged mining bee, *Dasypoda hirtipes*, the female has extraordinarily long, graceful drapes of hair arising from the outer surfaces of the hind tibiae, and greatly elongated hind basitarsi (Fig. 2.47). As we have seen, these are used to sweep away excavated sand or soil as the insect digs its burrow, but they also serve as very effective devices for pollen transport. There are also long hairs on the undersides of the femora, resembling those of *Andrena* species, though much less well-defined. Females of *Melitta* species are similarly equipped. As already mentioned, females of *Macropis europaea* collect a mixture of oils and pollen on their hind legs. The hind tibiae are clothed, on their outer surfaces, with a dense pile of long, whitish hairs, while the broadened basitarsi have a similar coating of long black hairs, with apical brushes of stiff hairs (Figs. 2.46, 5.21).

Females of *Anthophora* species have expanded and flattened hind tibiae and basitarsi that are densely clothed on the outer surface with long, backward-directed hairs. In the females of the familiar *Anthophora plumipes*, the hairs on the hind tibiae are bright orange, and give the superficial appearance of the full pollen loads of a small bumblebee (Fig. 2.65). At the apex of each basitarsus is a dense, square-ended brush of stiff hairs. The pollen-carrying equipment of *Eucera longicornis* is very similar, and it also has a stiff brush at the apex of each basitarsus. The basal segment of each fore tarsus also has a projection at its apex with a fan of stiff hairs, used in pollen collection.

The leaf-cutter and mason bees (*Megachile* and *Osmia* species) carry pollen among long, projecting but slightly backward-directed hairs on the undersides of the abdominal segments (sternites). The fore and mid tarsi are equipped with brushes of shorter, stiff hairs which are used both for grooming pollen from their bodies, and also for collecting pollen from flowers. Mid and fore legs are used to pass pollen back to the hind legs and from there it is packed into the pollen brush under the abdomen (Figs. 5.26, 5.27). The equipment of female *Anthidium manicatum* is very similar (Fig. 5.28), as is that of *Heriades truncorum*.

Fig. 5.23 A female *Andrena marginata*, showing feathery pollen hairs.

Fig. 5.24 A female *Andrena flavipes*, showing its pollen-carrying structures.

Fig. 5.25 A female *Lasioglossum malachurum* with full pollen loads on its hind legs.

Fig. 5.26 A female *Megachile* brushes its abdominal pollen hairs against the stigma of a *Geranium* flower.

Fig. 5.27 A female *Osmia* collects pollen on its abdominal pollen hairs.

Fig. 5.28 A female *Anthidium manicatum*, showing its abdominal pollen hairs.

Although there is diversity in the detail of the nectar- and pollen-collecting adaptations across different bee genera there is little that suggests an obvious adaptation to one or more floral syndromes. Species that collect pollen under their abdomens might seem to be adapted to collect pollen from flowers whose sexual parts brush against their undersides. Both *Osmia* and *Megachile* species are frequent visitors to the flag-type blossoms of the family Fabaceae and also to thistles and knapweeds of family Asteraceae, supporting this expectation, but they do also visit flowers of other types. *Anthidium manicatum*, with its similar collecting apparatus, is also promiscuous in the floral types it visits, and is a particularly frequent visitor to flowers of the deadnettle group, notably black horehound (*Ballota nigra*) (Fig. 5.28), which deposit their pollen from above. As mentioned above, tongue-length is frequently used to distinguish among pollinators in terms of their ability to access different floral structures in search of nectar. There is little doubt that the very long tongues of hawk-moths and butterflies give them access to some types of flower that would be inaccessible to shorter-tongued bees or flies. However, this has to be taken in relation to overall body size and shape, as well as the agility and learning ability of the insects concerned. Floral resources, especially if they are abundant, are rarely accessible in only one way.

Temperature, climate and foraging efficiency

Researchers such as Heinrich, who have drawn attention to the importance of energy budget for foraging bees, provide important insights. The geographical range of many species that occur in the British Isles is strongly constrained by climate. Southern and south-eastern England have by far the greatest number of resident species, and species richness declines northwards (see Baldock (2008) for a comparison of bee diversity across several English counties, and Edwards (2013b) for a guide to Scottish bees). Warm, sunny weather during their flight periods is important for many species, and, on the local scale, most species favour nesting sites that enjoy warm, sheltered microclimates, and many use walls, cliffs, banks or root plates with a south-facing aspect. The furry coats of many species are helpful in maintaining body temperature, and some species are able to raise their body temperature (especially their thoracic temperature) above ambient values by shivering their wing muscles. However, external sources of heat, such as heat radiated or conducted from warm soil or stones, and, more often, direct radiation from the sun, are frequently necessary for bees to bring their thoracic temperature up to that required for flight. Some species are also thought to warm up by resting on the petals of bell-shaped flowers, such as buttercup.

Within limits, the warmer it is the faster a bee can fly and the more rapidly it can forage (although bumblebees that are generally cold-adapted can become over-heated). As we saw in relation to the mate-searching behaviour of male *Anthophora plumipes*, reproductive success can depend on the number of hours through the day when the insects are able to be active, and their relative ability to be active in unfavourable weather. If anything, the pressures on female solitary bees are even greater. Many cleptoparasites gain entry to host brood cells during the time between the initial cell construction and the completion of the process of provisioning it, egg-laying and sealing the cell (see Chapter 4). For those species vulnerable to cleptoparasites (and most species appear to be plagued by at least one), there is thus a very high premium on speed of collection of the food mass for each cell, and ability to forage effectively in suboptimal conditions.

For bees adapted to relatively cool climates, those that can fly at low ambient temperatures, and/or rapidly raise their body temperatures above ambient levels, should have a reproductive advantage, and so should be selected for over evolutionary time. Stone and Willmer (1989) investigated the relationships between bees' ability to raise their temperature by muscular shivering (endothermy), the minimum temperature at which they were able to fly, the thermal regime to which they were adapted, and their body mass, for 55 species of bee. The relationships between these variables turned out to be complex, and had to be interpreted in the light of the family relationships among the bees (belonging to six families). Other factors being controlled for, bees that are adapted to cooler climates were able to fly at lower ambient temperatures, were able to warm up more rapidly and had higher thoracic temperatures in flight than those adapted to warmer climates. Controlling for family relationships and climatic adaptation, the researchers also found a strong positive correlation between size and warm-up rate. Within species, this suggests that larger individuals should be able to forage at lower temperatures than smaller ones – an expectation confirmed for the hairy-footed flower bee *Anthophora plumipes* (Stone, 1993). It would be an interesting exercise to relate these aspects of thermoregulatory ability in different bee species to their geographical distribution in the British Isles, and also to their response to climate change (see Chapter 6).

The territorial behaviour of male *Anthidium manicatum*, which can displace even bumblebees from its territory, is an obvious example of foraging behaviour affected by something other than the supposed fit between pollinator and floral syndrome. Stone (1995) also observed changes in foraging behaviour of female *Anthophora plumipes* faced by male sexual harassment. However, it is likely that variable features of bees, such as tongue length and thermoregulatory abilities, which relate to their foraging behaviour, must also affect the interactions between foragers of different species active at the same foraging sites.

Corbet and others (1995) provided a particularly clear way of representing the way these determinants of foraging behaviour might affect competitive relationships between bees foraging for nectar at different times and under different conditions. They focus on three main variables: species-specific costs of foraging (in effect, the size of the bee), the accessibility of the nectar (effectively, tongue length, considered in relation to the depth of the flower and the standing crop of nectar) and minimum ambient temperature for flight. The combination of these variables as they apply to particular bee species will specify thresholds below which they will not be able to forage profitably for nectar. Depending on the actual ambient temperature, the depth of available flowers and the quantities of accessible nectar in them, different species of bee will be advantaged or disadvantaged *vis-à-vis* others. For example, flowers are likely to contain more nectar early in the day, as there is less evaporation and less depletion from foragers, so bees that can forage at low ambient temperatures will be advantaged. Towards the middle of the day, when nectar crop is depleted, larger, long-tongued bees may monopolise deep flowers where depletion of their nectar takes the meniscus out of reach of shorter-tongued bees, but there is still enough to fuel continued foraging for longer-tongued ones. Alternatively, when competition between foragers for nectar from shallow flowers (e.g. the bowl- or dish-shaped syndrome described above) has depleted the nectar in them, smaller, shorter-tongued species may still be able to forage profitably as the nectar is replenished, but larger, long-tongued species may be excluded in virtue of their greater energetic costs of flying and maintaining body temperature.

Corbet *et al.* (1995) devised their 'competition box' as an aid to considering the potential effect on an existing pollination system of introducing a new species (usually honeybee or bumblebee) either to enhance pollination or for honey production. However, the concept could also be applied in analysing the interactions between different long- and short-tongued guilds of solitary bees as they forage for nectar at foraging sites with different mixes of wild and cultivated flowers. A significant feature of their analysis is that it draws attention to likely shifts in the species-composition of pollinator assemblages at a given site at different times of day (see Chapter 6).

5.4 Specialised mutualism or generalised interaction?

The competition box implies that foragers of each species may be more or less specialised in the flowers they visit, depending on the mix of other species foraging at the same patch, and on the time of day, temperature, and mix of flower-species. This takes us back to the question of how these complex patterns of flower visitation by bees of different species, each foraging for its own purposes, provide an effective pollination service for the plants. For cross-pollination to be effected, an individual forager must collect pollen from the anthers of a flower and carry it to the receptive stigma of a flower on another plant (the next one it visits?) of the same species. Although pollen-collecting females might seem well-suited for this, as they displace considerable quantities of pollen onto their body hair, they actively groom it onto concentrated loads, often moistening it with nectar as they do so. Might there be too little pollen left on the general body hairs of the bee by the time it reaches the next exposed stigma of the right species of plant? In fact, the grooming of body hairs by the fore and mid legs is not completely efficient, and, in any case, foraging bees usually visit several flowers in succession before grooming and packing the pollen that gets caught in their fur.

But there is still the matter of what chance there is that the next flower visited will be on a plant of the same species as the previous one. If the bee is foraging in a site dominated by a single species of concurrently flowering plant (such as, for example, a field of oilseed rape) this probability will of course be very high. However, in the natural and semi-natural habitats to which bees have become adapted, there will be a choice of flowers of different types with a variety of rewards in terms of availability, quality and quantity. In these settings there are two possible ways in which the foraging strategy of bees might incidentally favour the reproductive needs of the plants. One way would be the outcome of a prolonged coevolutionary process, in which particular pollinator species became specialised visitors to the flowers of particular plant species. For reasons to be discussed below, this sort of close mutualism is rarely found. However, as we will see, some degree of inherited specialisation on the part of solitary bees is relatively common.

The other way that bee behaviour might enhance pollination is known as flower constancy or fidelity. This is behaviour learned by individual bees, and it implies regular, successive foraging visits to flowers of the same species. This behaviour has been most extensively studied in bumblebees and honeybees, though there is some published evidence of it among solitary bees of several genera (e.g. Proctor, Yeo & Lack, 1996; Westrich & Schmidt, 1987; Minckley & Roulston, 2006). Since bees foraging with constancy in mixed arrays of flowers could well be missing more rewarding flowers of other species, this seems to be a sub-optimal foraging strategy. In their different ways, bumblebee and honeybee workers pool information about which flowers are currently offering high rewards. However, even these social bees still forage with constancy until prompted to switch to an alternative flower species, and, of course, solitary bees do not have the advantage of social communication.

So, why do so many individual bees show flower constancy, despite its evident disadvantages? The most favoured explanation is that switching from one flower-type to another would lead to a loss of foraging efficiency because of limitations in cognitive ability and motor-learning on the part of the bees. However, there is evidence that experienced bumblebees can learn to switch back and forth between two floral types without loss of foraging time, and Menzel (2001) offers an account of bee cognitive ability that tells against this explanation of constancy. An alternative view (derived from predator-prey relationships) is that the bee forages with a search image in its short-term memory. Experimental evidence seems equivocal, and Gegear & Laverty (2001) provide evidence that where bees forage in diverse plant communities with multiple variations of floral traits, this promotes forager selectivity, whatever the cause. If this is right, it suggests that the diversity of floral patterns do enhance pollination, but by incidentally promoting constancy, rather than by attracting specifically adapted pollinators.

But what of the alternative way in which plants might secure a reliable pollination service: inherited mutual specialisation between bee and plant? A complex set of floral traits might have coevolved with the sensory biases, physical structure, tongue-length and other characteristics of a partner-bee species. As mentioned above, such close ties at the level of particular species are rare. The reasons why such tight mutualisms are rare are partly related to the life-histories of bees. Social bees such as honeybees, bumblebees and the small number of halictid social species have long flight periods, greatly exceeding the flowering periods of any of the flowers they visit. Some solitary species are also on the wing for long periods, such as the small carpenter bee, *Ceratina cyanea*, which flies from April to September. Some other species, such as *Andrena flavipes*, have two or more generations in the year, and have distinct spring and summer broods. For all these species, fixed, inherited dependence on one plant species for all their nutritional requirements would be out of the question. They need to have the learning ability and flexibility to forage from a wide spectrum of flowers across the seasons.

Most solitary bees, however, are single-brooded in the British Isles, and they have relatively short flight periods. Perhaps these bees could have evolved close specialisations to forage from particular flower species? However, even for these bees, and the plants whose flowers they visit, it is a high-risk strategy to become dependent on the food sources or pollination services of a single partner. Especially in temperate climates, unpredictably adverse weather can lead to drastic falls in the population of an insect species, while others suffer cyclically from the ravages of predators and parasites. Variable weather patterns can also affect either the emergence times of the pollinator or the flowering period of its favoured forage plant, so that they might just miss each other. On the larger scale, while plant and pollinator may have overlapping geographical distributions, there may be climatic or other constraints that prevent one or other of the 'partners' from occupying parts of the distribution of the other. This situation is likely over evolutionary time to lead either bee or plant species to become less choosy.

These considerations point in the direction of the evolution of generalised, rather than specialised relationships between flowers and their pollinators. Studies at the level of whole pollinator/plant communities bear this out. Very few plant species have evolved floral morphologies that absolutely exclude visits from a large range of potential pollinators, while pollinating insects, including bees, show remarkable skill in accessing the rewards of a wide spectrum of flower types. However, within these very broad limits there is evidence that the foraging strategies of solitary bees do vary both among the different species, and in relation to different floral rewards. Most studies show that bees forage from a wider range of flowers when in search of nectar than when collecting plant oils or pollen. Since plant oils are an unusual food reward (and may also be used in nest construction), it is not surprising that the specialist association between yellow loosestrife and

Macropis europaea is very close (Fig. 5.21). This includes, as we have seen (Chapter 3), the use of the plant by males as the key to finding mates, as well as its use by females for collecting both oils and pollen. Significantly, other species of *Macropis* that occur in mainland Europe also specialise on plants of the loosestrife genus (*Lysimachia*) (Westrich, 1996). However, both males and females may take nectar from other flower groups (Edwards, 1998).

Despite the risks of dependency on a single plant for a necessary food source, a significant minority of solitary bees does appear to have evolved specialised preferences in foraging for pollen. In the most extreme case, where a bee species collects pollen from one species of flower only, it is described as monolectic. Where a species collects pollen from species in the same genus, sub-family or family only, it is said to be oligolectic. Those species whose females collect pollen from numerous families of flowering plants are described as polylectic. The red mason bee, *Osmia bicornis*, for example, collects pollen from as many as 19 plant families (Westrich, 1996). Estimates of the proportion of solitary bee species that fall into these three categories vary according to geographical area. In global terms, it seems that oligolectic species constitute a higher percentage of local bee assemblages in the tropics, and subtropics than at more northerly latitudes. According to Westrich (1989) oligolectic species make up about 30% of the bee fauna of northern and central Europe, while according to Pekkarinen (1997) they form only 15% to 20% of the bee fauna of Sweden and Finland.

Several species that occur in the British Isles are highly specialised in their foraging for pollen, and some have been regarded as monolectic. In addition to *Macropis europaea*, these include *Melitta tricincta, Colletes hederae* and *Andrena florea*. *Melitta tricincta* (Fig. 2.50) has a late flight period, from late July into September, synchronising with the flowering of red bartsia (*Odontites vernus*). Females appear to be monolectic on red bartsia in the British Isles, and both males and females forage predominantly from this plant, occasionally visiting other flowers for nectar. *Colletes hederae* (Fig. 5.34) is a formerly scarce species, currently expanding beyond its strongholds in southern and south-eastern coastal districts. Like *Melitta tricincta* it flies late in the year, synchronising with the flowering of ivy (*Hedera helix*) in September and October.

Andrena florea is a very localised mining bee of south-eastern England. The females collect pollen only from

Fig. 5.29 White bryony on an unmanaged hedgerow.

Fig. 5.30 Male flower of bryony.

Fig. 5.31 Female flower of bryony.

Fig. 5.32 A female *Andrena florea* with full pollen loads brushes against the stigmas of a female bryony flower.

Fig. 5.33 A male *Andrena florea*.

white bryony (*Bryonia dioica*) in England (Figs. 5.29–5.33). However, they also collect pollen from a closely related plant (*Bryonia alba*) in Germany (Edwards & Telfer, 2001). So, although this species is treated as monolectic in England, it seems that it is only so because closely related plants that it could use are simply not present. This may turn out to be true of most, or even all, supposedly monolectic species, and researchers tend to use the category of oligolectic bees to include the full range of more-or-less specialised pollen collectors. This lumps together species that have very different foraging strategies and makes it a large and diverse grouping, so Cane & Sipes (2006) have suggested a set of finer discriminations among the pollen specialists. Species, such as *Andrena florea,* that confine themselves to plants of a single genus are put together with others, such as *Andrena hattorfiana* (Fig. 2.36), a specialist on scabious (*Scabiosa* and *Knautia* species) and knapweeds (*Centaurea* species), and *Melitta haemorrhoidalis* (Fig. 2.48, a specialist on bellflowers, *Campanula* species) as 'narrowly oligolectic'. Species that take pollen from a single family, or, in the case of large families, a small number of genera or subfamilies within a family, are to be described as oligolectic, with a new term 'mesolectic' to describe species that gather pollen from flowers belonging to as many as three families, and are intermediate between oligolectic and polylectic species in their degree of choosiness in their pollen-collection. This terminology may not gain currency, but the point is clear: there is a continuous spectrum between extreme specialisation and extreme promiscuity among bee species in the plants from which they collect pollen.

Given the risks involved, the frequency of some degree of pollen specialisation among solitary bees is surprising. This raises the question: under what conditions might it have evolved, and continue to be sustained? Several hypotheses have been suggested, with some parallels to the discussion of learned dispositions to flower constancy. One possibility is that pollen specialists forage more efficiently on their preferred hosts than generalists using the same flowers. The evidence on this is not conclusive, and in any case has to be measured against the benefits to generalist foragers of being able to switch to more rewarding flowers if they are available. Another possibility is that mutual specialisation, in restricting the range of pollinators foraging from a particular flower or group of flowers, benefits bee specialists by reducing competition for floral resources. Unfortunately for this idea, it seems that pollen specialists tend to be associated with flowers that are attractive and popular with other pollinators, too. A third suggestion is that there are costs associated with evaluating and switching between different types of flower that have to do with limited cognitive and learning abilities, so for some species specialisation gives advantages that might outweigh the costs of being tied to a single floral type.

This discussion assumes that pollen specialisation, when it occurs, is a genetically inherited trait. However, some studies have investigated the role of learning in bees' choice of host plant flowers. Not only do inexperienced foragers have definite preferences, but these change with experience. Dobson (1987) carried out a series of experiments on the foraging behaviour of bees drawn from a population of an American species of *Colletes*. Though the species as a whole is not strictly oligolectic, bees in this population collected pollen exclusively from one plant species (*Grindesia stricta*). Dobson found that newly emerged (laboratory reared) bees showed a definite preference for *Grindesia* flowers, pollen and pollenkitt (the oily coating of pollen grains). Dobson attributed this set of preferences, and the associated ability to make chemical discriminations, to the pre-emergence larval experience of the bees. The hypothesis is that larvae become conditioned to the chemistry of the pollen, and especially pollenkitt, upon which they feed as larvae, and that this is expressed as an initial preference in the foraging behaviour of the adults.

The interplay of genetically coded preferences and learned dispositions is not fully understood, but a study of the evolutionary history of one genus of bees (*Andrena*) in the USA suggested that, over the long term, lineages switched from pollen generalist to specialist and back again quite frequently (Larkin *et al.*, 2007). This implies that foraging strategies are relatively flexible and open to modification in the face of selective pressures of the sorts just discussed.

Another possibility telling in favour of specialisation in some species might be that the preferred pollen has higher nutrient value, or, perhaps, suits the particular physiology of the larvae of that bee. Such studies as have been conducted seem to indicate that larvae artificially fed on the 'wrong' pollen do not seem to suffer and, in some cases, thrive on it. Also, although the nutrient value of the pollen from different sources varies considerably, there is no evidence bees can detect this. However, it is well-known that some plant chemistry includes toxins as a defence against herbivores, and that these toxins are sometimes also present in pollen. An evolved specialisation to collect pollen from a highly rewarding flower might well confer an evolutionary advantage for a bee with larval immunity to its toxins. A study of 60 species of the genus *Colletes* that occur in the western Palaearctic (Müller & Kuhlmann, 2008) compared pollen-generalist species with specialist ones. Fourteen pollen-specialists collected pollen predominantly or exclusively from one sub-family of the family Asteraceae (the very diverse Asteroideae), while it was very little used by most of the generalists. The pollen of this group of plants contains protective compounds that render it difficult to digest, suggesting that the larvae of the pollen-specialists may be physiologically adapted to utilise it. A further study by Sedivy *et al.* (2011) involved rearing larvae of two closely related pollen-generalist *Osmia* species on restricted diets. There were marked differences in the ability of the larvae of the two species to survive on diets consisting of the same pollen, strongly suggesting that digestion of different pollens requires physiological adaptation.

Minckley and Roulston (2006) give evidence from their own and other studies of pollination systems that specialist bees, often of several different species, frequently forage from the same dominant, reliable and highly rewarding flowers as do generalists. They propose that specialisation among pollinators is associated with specialisation to fly synchronously with such abundant and predictable host plants. This suggestion fits well with at least two of the highly specialised bees mentioned above. Both *Colletes hederae* and *Melitta tricincta* (Fig. 2.50) fly late in the season when there are relatively few reliable forage plants (in natural or semi-natural habitats). Ivy, especially, is notable as a food bank for a great range of bees, wasps, flies and other invertebrates on the wing when few other flowers are available.

Ideas such as the pollination syndrome and the most effective pollinator focus attention on interactions between particular pollinators and plants, but ecologically informed approaches to pollination have recognised that at the level of local communities there are multiple interactions between flowers and their pollinators, some of them highly specialised, others more generalised in both directions. In recent decades, a combination of increased computing power and ecological mathematical modelling has produced approaches to the study of pollination systems as complex networks that have structural properties not otherwise detectable (Jordano *et al.*, 2006). These show, for example, that total observed interactions in a pollination network have a non-random pattern, with a definite structure of nested relationships. This confirms the intuitive sense that there are real associations between different floral types and their pollinators, but they do not commonly take the form of tight species-to-species interdependencies. Where bees gather resources, notably pollen, from a limited range of flowers they usually do so alongside numerous other flower-visitors, at least some of which may be effective pollinators.

Fig. 5.34 A female *Colletes hederae* collects pollen from ivy blossom.

Fig. 5.35 A female *Colletes halophilus* collects pollen from sea aster.

Fig. 5.37 A female *Chelostoma campanularum*, a bell-flower specialist, collects pollen from the immature female part of a cultivated *Campanula* flower, after the stamens have withered.

Fig. 5.36 A female *Colletes succinctus*, a heather-specialist, collects pollen from ling.

As we have seen, this is true of *Colletes hederae* and ivy. Other narrowly oligolectic bees, such as *Andrena florea*, *Andrena fuscipes* (a heather-specialist) (Fig. 5.19), *Chelostoma campanularum* (Fig. 5.37) and *Melitta haemorrhoidalis* (Fig. 2.48), also share their favoured flowers with numerous other visitors. Asymmetrical mutualisms of this sort seem to be a standard feature of pollination networks, and understanding them can be important for effective conservation of both plants and their pollinators.

6 The conservation of solitary bees

6.1 Introduction

In recent decades researchers and activists alike have drawn public attention to a growing 'pollination crisis' (Buckmann & Nabhan, 1996; Kearns et al., 1998; Friends of the Earth, e.g. press statement on 6 March 2014). Many flowering plants, and especially those used as staple crops across large parts of the world, are pollinated abiotically, by wind or water. However, these crop species make up only a tiny proportion of flowering plant diversity, and the great majority of species depend on animals to pollinate them. Of these, some are self-pollinated, but need animals to transfer pollen from anther to stigma. Others may be self- or cross-pollinated, but seed-set is generally much higher (by up to 43%) when they are cross-pollinated. Still other plants are self-incompatible and so completely dependent on cross-pollination. As we saw in Chapter 5, a recent estimate gives an overall total of 87.5% of flowering plant species globally that are animal-pollinated, with slightly fewer (78%) in temperate zones. Of plants grown for food, some 75% either depend on or are more productive with, animal pollination. Some authors provide estimates of the economic value of animal (in effect, insect) pollination as just under 10% of the economic value of all agricultural production (Gallai et al., 2009). Up to one third of human food consumption is derived from insect-pollinated crops, and demand for these crops has grown three-fold since 1961 (Aizen & Harder, 2009).

One of the reasons for alarm has been the widely publicised threat to honeybee populations. As many honeybee colonies fell victim to *Varroa* mite and so-called colony collapse disorder, it became clear that reliance for most crop pollination on just one commercially managed species of pollinator was a high-risk strategy. Some crops, notably tomatoes, have also been pollinated in enclosed beds by managed bumblebees, usually *Bombus terrestris* as, unlike honeybees, these can buzz-pollinate: that is, produce high-frequency vibrations that cause the anthers of the tomatoes to release pollen (Corbet & Huang, 2014). It has long been recognised that wild bumblebees also provide an important 'backup-service' in pollinating orchard crops as well as field beans and other Fabaceae. Unfortunately, from the 1980s onwards several sources of evidence pointed to very severe declines in the distribution of many bumblebee species, both in the UK and other parts of Europe (Williams, 1982; Westrich, 1989; Benton, 2006).

Given the manifest problems associated with the best-known pollinator, as well as its most thoroughly studied alternative group, the potential importance of other, neglected, wild pollinators came into focus. At the global level, the parties to the UN Convention on Biological Diversity set up an International Initiative for the Conservation and Sustainable Use of Pollinators at their meeting in 2000, inviting the UN's Food and Agriculture Organisation to coordinate the Initiative. The intention was to monitor pollinator decline and any impact on crop production, and to evaluate the economic importance of pollinators, as well as to improve taxonomic knowledge and promote pollinator diversity. This was followed by a number of regional initiatives, including the UK's Insect Pollinators Initiative, which brought together research councils, Defra, the Wellcome Trust and Scottish Government.

The focus on the importance of bees, in particular, as pollinators of agricultural and horticultural crops, and attempts to set an economic value on the 'ecosystem services' thus provided have certainly brought welcome political attention and research money to the conservation of wild bees. However, this strategy for conservationists is a risky one – especially if it is pursued as the only, or even the main one. What if it is shown that species redundancy in generalised pollination webs makes conservation measures unnecessary, or that alternative plant-breeding methods, or new technologies might allow us to dispense with the pollination services of wild bees? Might we not move towards a situation in which only those species known to have an economic pay-off or demonstrated human utility are deemed worthy of conservation? True, many of those who foreground the threat to crops also write of the parallel threat to wild plants from lack of suitable pollinators, but that is not where the research effort is focused.

An alternative approach to bee conservation might take the evidence of their decline, especially in agricultural landscapes, as just one symptom among many of a wider malaise in currently dominant patterns of land use and methods of food production, distribution and consumption. If these turn out to be unsustainable, then to base conservation on units of measurement appropriate to our current system may miss the chance to raise deeper questions about our relationship to the rest of nature. Highlighting our dependence on the 'ecosystem services' provided by the diversity of living species makes an important contribution to the case for conservation. However, an approach which goes beyond that, enjoining us to see ourselves as part of an immensely complex and intertwined community of species and relations, in which other species are valued for their own sake, and for the wonder and fascination of the opportunity to encounter them, could be still more powerful. It is arguable that Rachel Carson's pioneering *Silent Spring* (1962) was a major inspiration for path-breaking environmental legislation in the USA and elsewhere. It was an important influence on the emergent environmental movement and remains one of the most highly regarded works of non-fiction of the 20th century, more than 50 years after its publication. Perhaps its colossal impact had more to do with public horror evoked by the prospect of a world without birds and bees than concern with the economics of agriculture? A recent publication by a large international group of signatories (Kleijn *et al.*, 2015) adds weight to the case for supplementing economic arguments for bee conservation with a moral claim. Bees which currently predominate in the pollination of agricultural crops, and whose conservation is most cost-effective, usually belong to a small number of common species, and are not of high conservation priority. This is not surprising, as they are ones that have survived despite several decades of intensification and habitat loss. If we were to depend solely on economic arguments in terms of ecosystem services, then the great majority of species, including ones that are threatened and in most need of conservation measures, would be entirely without protection.

6.2 Are bees in decline?

As we saw above, declines in honeybees and bumblebees have been well-documented. The status of other wild bees, those discussed in this book, is harder to establish. There are no internationally standardised methods of monitoring them, and the difficulty of identifying many species in the field has limited the development of citizen science initiatives such as those that have benefitted bumblebee and butterfly conservation. However, there is telling evidence of decline, especially in agricultural landscapes, across large parts of Europe and America. A study by Biesmeijer *et al.* (2006) took data from surveys in Britain and the Netherlands, comparing local bee and hoverfly diversity before and after 1980. The results were complex, but showed substantial declines in bee diversity in 52% of the 10 km squares in Britain, with increasingly uniform bee assemblages, and relative dominance of a small number of generalist species in the post 1980 data. The study also showed greater declines of oligolectic bees than polylectic ones, and associated declines in more specialist plant species. Buckmann & Nabhan (1996) drew on comparable evidence for parts of the USA. Reviews of evidence of declining bee diversity in Britain include Falk (1991), O'Toole (1993), and, for Ireland, Fitzpatrick *et al*. (2007).

However, not everyone has been convinced that there really is a pollination crisis. Gazoul (2005) argued that dramatic claims might be misplaced, or, at least, premature. Most staple crops do not depend on insect pollination, while ones that do are often grown in smaller scale systems which tend to be low-input and consistent with high diversity of wild pollinators. While Gazoul conceded that some notoriously intensive systems in the USA suffer by over-stretching pollinator services, he argued that evidence of bee and butterfly declines in Europe relate to rare and specialised species. Generalist flower-visitors remain common, and have even expanded their range, and it is these generalists that provide most of the pollination of crop plants. So far as wild plant species are concerned, the degree of generalisation and asymmetry in pollination webs makes it unlikely that loss of particular pollinators would have much effect on the reproductive success of these plants. However, even Gazoul conceded that environmental changes are often not recognised until too late, and advocated suitable conservation measures: retention of natural and semi-natural habitat for wild pollinators in agricultural landscapes, domestication of suitable solitary bees, and long-term monitoring of pollinator populations.

However, Steffan-Dewenter *et al*. (2005) were able to cite evidence of bee decline in both the USA and Europe not mentioned by Gazoul. They also draw on the standard work on crop pollination (Free 1993) and more recent sources in support of the view that even generalist crop plants do benefit from species-diverse pollinator assemblages. They cite evidence that in intensive agricultural landscapes there is a decline in both abundance and diversity of pollinating insects associated with distance from natural or semi-natural habitat. On the dietary significance of insect-pollinated crops they argue that Gazoul pays insufficient attention to the nutritional value of the proteins, vitamins and minerals that are obtained from them, and to the desirability of dietary diversity. Gazoul's claim that most insect-pollinated crops are grown in small-scale and high diversity systems is rapidly becoming outdated with the spread of intensive systems.

Since the time of this exchange, further evidence, including the above study by Biesmeijer *et al.* (2006), has come in from numerous studies, all pointing to widespread grounds for alarm. Potts *et al.* (2010) review much of this evidence, including studies of extensive pollen limitation (due to inadequate pollination) as impairing the reproductive processes of wild plants, as well as a growing disparity between the expanding production of insect-pollinated crops and the availability of domesticated pollinators. The evidence is that many commercial systems depend on the unacknowledged 'free' pollination services of threatened wild pollinating insects.

Of particular interest are the implications of increased knowledge of the structure of pollination webs. Given the high proportion of generalists in most systems, and the tendency for specialist plants to be pollinated by generalist pollinators and *vice versa*, it seems likely that pollination webs should be very resilient in the face of loss of component species. A study by Memmott *et al.* (2004) simulated loss of generalist species, specialist species and random losses from two well-studied US pollination webs. The results, confirmed subsequently by other studies, were that the webs were, indeed, relatively robust. Species-loss did lead to declines in pollinator and plant diversity, but these were gradual, rather than catastrophic. For these systems, honeybees, bumblebees and members of several families of solitary bees were found to have the widest connectivity and also to be the species whose loss would have the greatest impact. However, the warning from this study is that the relative tolerance of pollination webs to extinctions of particular species should not be regarded as immunity. Complacency in the face of local extinctions should be avoided. Their data suggest that management favouring generalist wild pollinators would be the most appropriate way of avoiding threats to pollination services. More recently, Potts *et al.* (2010) argue that ongoing global changes may result in more severe disruptions of pollination webs that could reach tipping points and subsequently collapse.

6.3 Bee decline in agricultural landscapes

The most widely cited driver of the decline of bee diversity and abundance is the loss and fragmentation of natural and semi-natural habitat resulting from agricultural intensification. Efficient use of machinery has required destruction of hedgerows, drainage of wetland, and levelling off of landscape features. The resulting changes to agricultural landscapes include the isolation of habitat fragments, reduced size of such habitat patches, reduced population sizes and reduced densities of both plants and pollinators (Figs. 6.1–6.6). It is difficult to distinguish empirically between the effects of these different aspects of change on pollinators, but their combined effects are clear. Kearns *et al.* (1998) review a wide range of international studies showing deleterious effects of habitat fragmentation. Steffan-Dewenter *et al.* (2006) also report decline of pollinator diversity as a result of habitat fragmentation and land-use intensification. Solitary bees are particularly threatened by these changes, and, among them, the more specialised (oligolectic) species and cleptoparasites are especially vulnerable. A probable explanation for the greater vulnerability of solitary bees to habitat fragmentation (compared with, for example, honeybees or bumblebees) is the tendency for them to have relatively small home ranges (Gathmann & Tscharntke, 2002). As we have seen (Chapter 4) many solitary bees are also constrained by their vulnerability to invasion of their unfinished brood cells by cuckoos. Bees are dependent on appropriate sources of nectar (or plant oils) and pollen, appropriate nest-sites, and sources of materials for nest-building. Moreover these partial habitats (Westrich, 1996) must be within the foraging range of the bee. Westrich gives the example of *Colletes hederae* which was discovered nesting on a steep, sparsely vegetated slope in Germany. Its sole pollen source, ivy, was located on a wall in a nearby village – so the bee was vulnerable to destruction of either of these two, seemingly unconnected, resources.

Much less obvious, but potentially no less damaging, are the effects of isolation on the genetic structure of populations. Where patches of habitat are separated by large distances, or where there is loss of connectivity between them, relatively sedentary species may form dispersed small, inbreeding populations with no interchange between them. Davis *et al.* (2010) illustrate some of these processes in the rare solitary bee *Colletes floralis*. Such isolated populations tend to lose genetic diversity and become differentiated from one another, and are subject to random local extinctions (Ellis *et al.*, 2006).

Fig. 6.1 Intensive arable cultivation of former chalk downland, north-west Essex.

Fig. 6.3 Former coastal grazing marsh, now converted to intensive arable cultivation.

Fig. 6.2 Corn marigold and cornflower, bee-friendly arable weeds now eliminated in intensive regimes

Fig. 6.4 Newly ploughed coastal grassland, with remnants of semi-natural vegetation in foreground and far distance.

Fig. 6.5 Encroachment of intensive arable cultivation on South Downs hillsides.

Fig. 6.6 'Improved' pasture with grazing sheep. Spring forage for bees eliminated except for remaining hedgerow shrubs.

Surprisingly, Davis *et al.* (2010) found that urban settlements constituted more of an obstacle to gene-flow than agricultural land. However, the regions of north-west Scotland and Ireland where *C. floralis* occurs feature more extensive and varied agricultural ecosystems than are found over much of lowland Britain.

So far as wild plant species are concerned, smaller, isolated populations may suffer pollination limitation because of fewer visits, visits by the 'wrong' species, or absence of appropriate pollinators. Plants capable of self-fertilisation may persist, but suffer ill-effects of inbreeding.

In addition to fragmenting and isolating natural and semi-natural habitat patches, intensive agriculture presents other potential threats to bees and other pollinators. The most thoroughly researched (and still controversial) of these is the use of pesticides. Surveys of wild bees and butterflies have shown reductions in diversity where accumulated pesticide loads are high, and attention has more recently been focused on sublethal exposure to pesticides, most notably neonicotinoids. These include inability of foragers to relocate their nests, and reductions of learning ability and foraging performance, growth rate and colony productivity (in the case of social bees) (see Vanbergen *et al.*, 2013 for references). The use of herbicides, too, in frequently reducing both the abundance and diversity of wild flowers in remaining patches of foraging habitat adjacent to agricultural land, has negative effects on the nutrition of bees, especially oligolectic species. The use of nitrogenous fertiliser adds to this effect by promoting the growth of vigorous grass species at the expense of wild flowers. A

Fig. 6.7 Flowers in a roadside verge.

recent study conducted by Hanley & Wilkins (2015) in south-west England found that bumblebees preferentially foraged on roadside verges rather than on field-facing sides of hedgerows, where key forage plants were missing. It seems very likely that solitary bees would show the same behavioural pattern (Fig. 6.7).

The accidental or deliberate introduction of alien species of pollinators (as, for example, the use of managed honeybees to boost horticultural or agricultural production) to an already existing assemblage of wild species may have unintended effects in changing the community structure, out-competing native species and/or introducing pathogens (Corbet, 1996). The arrival of reward-rich alien plants could also be a cause for concern. Although there is little or no evidence for it, there is a theoretical possibility that they may co-opt important pollinators, attracting them away from native plants (Vanbergen *et al.*, 2013).

Each of the above challenges to bees and other pollinating insects may, or can be shown to, have adverse effects on abundance and diversity, but it is important to recognise that when they act simultaneously their combined impact is intensified. For example, increased parasite loads impose higher nutritional demands on bees that simultaneously confront more demanding foraging conditions. Conversely, reduced foraging efficiency due to fragmentation of resources or pesticide-induced disorientation may have a negative effect on immunity to disease and parasitism (Vanbergen *et al.*, 2013).

Finally, it might be thought that some agricultural crops offer benefits to wild pollinators in the form of highly rewarding supplies of pollen or nectar. Oilseed rape is one such crop that has been shown to enhance pollinator densities in some contexts (Westphal *et al.*, 2003) (Figs. 6.8, 6.9). However, it offers its bonanza for a short period only, so in large-scale monocultures it does not meet the requirements of species whose flight-period does not coincide with the flowering period of the crop, or which have prolonged flight seasons. Of course, only generalist pollinators can benefit, and the crop is frequently grown in uniform agricultural landscapes that lack provision for nesting habitat. Spring-flowering fruit trees, on the other hand, can provide nutritional rewards at a time of year when wild sources are very limited and, when these are managed in low input and high diversity orchards, or in informal green spaces, where both nesting habitat and later-flowering plants are routinely available (Fig. 6.10).

Fig. 6.8 Oilseed rape in a landscape of intensive arable cultivation.

Fig. 6.9 A field of oilseed rape at the edge of ancient woodland in a complex landscape.

Fig. 6.10 Apple blossom providing spring forage for bees in an urban green space.

Conservation strategies for agricultural landscapes

One widely advocated strategy for addressing the risk involved in dependence on just one managed pollinator is to develop methods of commercially managing other species of bee. The commercial management of *Megachile rotundata* as a pollinator of alfalfa crops is well established in the USA and elsewhere, while some species of *Megachile, Osmia* and *Xylocopa* have been advocated as potential pollinators for crops such as Solanaceae (including tomato) and *Vaccinium* which require buzz pollination. A number of solitary bees are increasingly used for pollination of tropical fruit crops, while in parts of Europe, Asia and North America several species of *Osmia* are used for pollination of spring-flowering orchard crops such as apples, almonds, cherries, pears and plums (Williams, 1996; O'Toole, 2012, 2013). There is little doubt that, at least in the short term, management of selected solitary bee species may offer alternatives to dependence on honeybees and one or two bumblebee species. Indeed, O'Toole (2013) cites the horn-faced mason bee, *Osmia cornifrons*, as 80 times more effective than honeybees as a pollinator of apples. However, there are grounds for caution. Corbet (1996), for example, points to the risk of disruption of already-existing assemblages of pollinating insects. As a strategy for dealing with advancing problems of pollination limitation in many crop plants it could be argued that it serves only to provide short-term respite while delaying the more fundamental reform of intensive systems that is needed. Of course, as a strategy for dealing with the decline in abundance and diversity of bees and other insects, it has very little indeed to offer.

The most widely advocated alternative, which does offer real prospects for bee conservation, while also serving crop productivity and economic viability, is to enhance the agricultural landscape. After relatively unpromising beginnings, agricultural set-aside and environmental stewardship schemes have increasingly included options favourable to wild flowers and their insect visitors (Figs. 6.11, 6.12). Several studies (reviewed in Dicks *et al.*, 2010) of the effects of field margins sown with nectar and pollen mixes of either wild or agricultural flowering plants reported increased numbers and diversity of bumblebees. It seems likely that solitary bees, too, would have benefitted, but hard evidence is missing. Carrie *et al.* (2012) compared the attractiveness to 'beneficial' insects of a range of wild flowers, showing that this was highly variable in ways that might not have been predicted, and that an unintended consequence of some conservation measures might be enhancement of the populations of pest species. In some agricultural areas, hedges and patches of scrub offer more floral resources than herbaceous plants, especially in spring (Nick Owens, personal communication). Consistently with this, Banaszak (1996) argues for a more diverse agricultural landscape, including refugia for wild flora and fauna adjacent to meadows and cropped fields. Refugia include roadside verges, hedges, shelter belts and remaining areas

Fig. 6.11 Agricultural set-aside with limited value for pollinating insects.

Fig. 6.12 A field-margin sown with red clover and other bee-friendly plants.

of semi-natural habitat that should amount to not less than 25% of the landscape. Some studies confirm the intuitive expectation that bee diversity will be higher where there are areas of natural or semi-natural habitat close to crops or when, as Banaszak proposed, the proportion of such habitats in the overall landscape is high (Ricketts *et al.*, 2008; Stephan-Dewenter *et al.*, 2002**)**. Connectivity between areas of natural or semi-natural habitat may also be important, given evidence that small, isolated populations are liable to suffer reduced genetic diversity and increased vulnerability to local extinction. While most studies focus on bumblebees and treat floral resources as key to conservation, there is so far little systematic research on attempts to preserve or create nesting habitat for solitary bees in agricultural landscapes.

Although much can be done within conventionally farmed landscapes to render them less bee-hostile, a more fundamental and long-term sustainable approach would be to encourage modal shifts to more extensive, low input agricultural systems with biological pest control. There is evidence that these can be made as productive (in terms of crop yield per hectare) as intensive systems, but with much less environmental damage (see, for example, Kearns *et al.*, 1996). Where chemical pest control is attempted, current methods of risk assessment for pesticides that focus on lethal doses for honeybees need to be extended to include other bee species, and also take into account sublethal effects on physiology and behaviour (Vanbergen *et al.*, 2013).

6.4 Bees in non-agricultural landscapes

Although agriculture is the dominant land-use in Britain and Ireland, especially in lowland areas, there remain marginal tracts of land that have escaped agricultural intensification and others that so far are intractable for any kind of agricultural development (Figs. 6.13–6.18). Such areas of natural or semi-natural habitat have survived for a mixture of historical, economic, technical, political and geographical reasons. Some, such as sea cliffs and coastal dunes, are largely resistant to agriculture for technical reasons, while steep downland slopes and some wetland areas have resisted intensification for economic reasons. Other areas, such as ancient woodland, common land, and countryside of particular historic, aesthetic or ecological value may owe their persistence to campaigning social movements and political action to set them aside from the economic forces driving intensification. One form or

Fig. 6.13 Protected flower-rich machair grassland in the Hebrides.

Fig. 6.14 Unimproved chalk grassland on Salisbury plain, protected from intensification by military use

Fig. 6.15 A mixed urban-edge landscape with rabbit-grazed acid grassland, woodland edge, scrub, hay meadow, river-bank, wetland and arable cultivation.

Fig. 6.16 A mixed-use landscape on the lower Thames estuary, with unimproved pasture, a power station, old ditches, hedgerows and a historic monument.

Fig. 6.17 A nearby grazing marsh obliterated by landfill.

Fig. 6.18 Destruction of a Dorset heathland by housing development.

another of legal designation is usually needed to achieve this. In yet other cases, semi-natural habitat persists as an unintended consequence of such conditions as neglect, mixed land-use, planning blight, toxic residues from former industrial uses and uncertain ownership rights. The edges of urban settlements often share some of these 'edgeland' characteristics.

But agricultural intensification and mechanisation

have been intimately intertwined with industrialisation and urbanisation. Urban settlement, industrial plant and associated material infrastructures (road and rail links, flood defences and so on) have steadily encroached on both agricultural and amenity land uses. However, the dwellers in urban settlements have come to demand fresh air and access to nature – for garden cities, open spaces, parks and gardens. Industry itself, while encroaching on rural landscapes, has also produced incidental habitat in the form of worked-out quarries, so-called brown-field sites, waste-dumps and marginal habitats alongside road and rail tracks and on flood defences. The value of these for conservation of wildlife, and for bees in particular, is only partially recognised. Perhaps the most effective compensation for the decline of bees in agricultural landscapes will be a growing movement to protect and secure these often despised places (Fig. 6.19).

Coastal habitats, species-rich grassland, heaths and moors, stream and river valleys, fens and marshes persist, often as isolated fragments in wider agricultural or urban landscapes, where they may be protected by some legal designation or by their incorporation into a reserve owned or managed by a conservation organisation. Increasingly, conservationists are recognising the limitations of this approach, especially if surrounding land is managed in wildlife-hostile ways. Nowadays the watchword is 'landscapes', within which diverse areas of conservation value are linked together to enhance flows and connectivity. On the largest scale, national parks offer opportunities to realise such a vision, though along with scale comes

Fig. 6.19 Purple toadflax growing below the platform of a rural station, with nesting habitat provided by an old wall.

diversity of conflicting interests and perspectives. At smaller scales, country parks, Local Authority conservation zones and 'living landscape' projects initiated by Wildlife Trusts all offer some potential for management for wildlife at a landscape scale (Figs. 6.20–6.24). However, while it is important to work on the larger scale, this must not be at the expense of protection and appropriate management of often very small-scale micro-habitats that may be crucial for many solitary bees.

Perhaps the most important requirement for protection or management of any site for bees is to find out which species are there already. As we saw in Chapter 2, this is no easy feat. However, the publication of the first comprehensive field guide to the bees of the British Isles in over a century (Falk & Lewington, 2015, a worthy successor to Saunders, 1896) constitutes a great step forward. This book includes user-friendly keys to species, and a wealth of fine paintings and photographic illustrations. The reader can also access the author's web feature on Flickr (Steven Falk http://tinyurl.com/nrywslu). Even equipped with this resource, the large number of species that are very similar in appearance, and the need to distinguish microscopically fine anatomical details will be found challenging by many beginners. This handbook will, I hope, provide a useful stepping stone, and further help can be obtained in many parts of the country by joining a local natural history society, field club or Wildlife Trust. Many of these will have at least one local expert who can offer help with equipment, confirmation of your provisional identification, and provision of spare specimens for comparison.

Meanwhile, much useful information can be gained from scanning key features of a site. Flat, sloping or vertical south-facing exposures of bare or sparsely vegetated ground may be used by mining bees, and areas of bramble or gorse scrub, dead wood, old fence-posts and old, crumbling walls may all be valuable as nesting habitat for stem- or cavity-nesters. The floral resources present on the site and in the vicinity can be audited, noting seasonal availability and diversity of flower-types (Figs. 6.25–6.26). This should give some idea of the range of generalist bee species likely to be present. To take account of the possibility a site might hold one or more of the more localised pollen specialists, to which some management should be focused, care should be taken to note the presence of relevant host plants. These plants include white bryony, bellflowers, scabious species, red bartsia, heather, mignonette, weld, germander speedwell,

Fig. 6.20 A flower-rich hay meadow nature reserve, converted from arable by a local council.

Fig. 6.21 An informal urban green space, protected for its historical significance.

Fig. 6.22 An important archaeological site sown with wild-flower seed-mix and managed in a bee-friendly way.

Fig. 6.23 An urban green space with sallows and red deadnettles, providing spring forage for bees.

Fig. 6.24 Hadleigh Downs Country Park, painted by Constable, and home for numerous scarce bees.

yellow loosestrife, small Fabaceae, such as bird's-foot trefoil, and yellow Asteraceae, such as hawkweeds and ragworts. Any planned management should take account of which of these characteristics a site possesses, and, at least, avoid damaging or removing any of them. In the case of ragwort, a valuable bee-forage plant, land managers must take into account prevailing legislation and codes of practice. As a list of bee species present is developed, then specific measures might be taken to enhance the site for any species present of high conservation priority, or to generally increase the capacity of the site for bee species. These might include

Fig. 6.25 A female *Andrena clarkella* collects pollen from sallow.

Fig. 6.26 A female *Andrena haemorrhoa* forages from sallow in western Scotland.

Fig. 6.27 A bee hotel in an urban park.

adding bee hotels for stem- and cavity-nesters (Fig. 6.27), and clearing areas of vegetation or constructing south-facing banks for mining bees. In parks and gardens, especially, there are opportunities for introducing bee-friendly plants (see below). The effects of any changes should be carefully monitored by regular surveys, while bearing in mind that bee populations are liable to vary considerably from one year to the next independently of any change to their habitat.

Beyond these very general considerations, a few more specific points can be made about particular types of habitat that are likely to harbour significant numbers of bee species.

Coastal cliffs and sand dunes

These can be very rich indeed in solitary bees, and often need little or no active management (Figs. 6.28–6.30). Vertical or sloping cliff-faces provide substrates for nesting mining bees, including several *Colletes* species, *Dasypoda hirtipes* and *Andrena* species, with sparsely vegetated older slopes and cliff-tops often having flower-rich grassland. *Andrena thoracica* nests in sandy cliffs, mainly in southern and western coasts of England and Wales. Depending on aspect and latitude, sand dunes harbour some of our scarcest bees, such as *Colletes hederae*, *Colletes halophilus* and *Colletes marginatus* (mainly southern or south-eastern coasts of England) and *Colletes cunicularius* (mainly western coasts of England and Wales, though increasingly found inland). *Anthophora bimaculata* and *Megachile leachella* are widespread along southern British coasts. The edges of compacted footpaths are often favoured by large aggregations of *Lasioglossum malachurum*. Flowers such as sea aster and sea holly are attractive nectar and pollen sources for these species. Cleptoparasites such as *Epeolus variegatus* and various *Nomada* species will often be found where their host species occur.

Coastal flood defences, roadside verges, power lines and railway cuttings and embankments

These humanly created landscape features all have primary functions other than conservation, but, depending on their management, they can make important contributions (Figs. 6.31–6.34). The fragmentation and intensive use imposed on agricultural land does not apply to these infrastructures, so there is more freedom to manage them sympathetically for wildlife if the agencies concerned have a will to do so. Considerable headway has been made along some parts of the coast to include wildlife conservation as a management con-

Fig. 6.28 Cliffs near Hastings, nesting habitat for *Colletes hederae*.

Fig. 6.29 Essex coastal dunes, habitat for *Megachile leachella, Colletes halophilus, Dasypoda hirtipes* and many other species.

Fig. 6.30 Coastal grassland with sandy exposures, habitat for *Magachile leachella, Anthophora bimaculata, Eucera longicornis* and other species.

sideration on sea walls (Gardiner, 2012; Gardiner *et al.*, 2015), and the combination of flat tops, often with bare, trampled ground, sloping sides with varying aspects, and an inner, flat, 'folding' that is frequently flower-rich provide good nesting and foraging habitat for bees. The only remaining known Essex habitat for the declining long-horned bee (*Eucera longicornis*) is along a stretch of sea wall, but the habitat is suitable for a wide variety of other species too, especially mining bees such as *Lasioglossum malachurum*. Roadside verges can provide valuable habitat, especially if they include south-facing banks, and the same applies to railway cuttings and embankments. For each of these types of habitat, however, conservation value depends very much on the management regime. Many road and rail infrastructures are either not managed for long periods or subject to frequent mechanical mowing. As with many open grassland habitats, the prevailing view among bee conservationists is that some form of rotational management is preferable to either of these, so that a diverse mosaic of habitat patches is maintained. Dicks *et al.* (2010) cite evidence from the USA on the benefits of appropriate management of land under power lines as well as roadside verges.

Fig. 6.31 Flood defences between caravan parks, with nesting habitat on top and forage along slopes and folding.

Fig. 6.32 Bee-friendly flora along flood defences on the lower Thames estuary.

Fig. 6.33 Coastal habitat for *Megachile circumcincta*, south-east Scotland.

Fig. 6.34 A former railway cutting, with a rich chalk flora.

Heather heaths

These dry, sandy habitats support more bee species than any other in Britain (Figs. 6.35, 6.36). This is partly due to the relatively sparse vegetation, the tendency for the sandy soil to heat up quickly in the sun, and the ease with which burrows can be dug. These features are shared with coastal dunes and mineral extraction sites, and so there is considerable overlap in the assemblages found in all three sorts of site. Many of these species forage on flowers not found exclusively on heathland, and may visit a range of flowers on the edge of the heath, or in local gardens. Others make use of patches of gorse or bramble, for nesting, foraging or over-wintering. However, a few species fly from mid to late summer and are oligolectic on heathers. These include *Colletes succinctus*, *Andrena fuscipes* and *Andrena argentata*. Where these species occur their cuckoos, *Epeolus cruciger*, *Nomada rufipes* and *Nomada baccata* can often be found.

Much lowland heath has been lost to urbanisation, and many remaining areas are disconnected fragments or have reverted to scrub or secondary woodland because of lack of management. Urgent action is required to secure

Fig. 6.35 A Suffolk heath, habitat for many species, including heather-specialists such as *Colletes succinctus*.

Fig. 6.36 Steep exposure caused by vehicular movement in a Norfolk heath, providing nesting habitat for mining bees.

such remaining heaths as are not fully protected, and to implement appropriate management (Fig. 6.37). Measures to restrict vegetation succession might include some localised grazing or controlled use of fire, but disturbance by heavy vehicles, including deep grooves in the sand made by vehicle tracks, is also very effective in maintaining nesting habitat. Bramble provides useful spring and summer forage, as well as nesting and hibernation habitat for stem-nesting species, such as the scarce bee of south-eastern England, the small carpenter bee (*Ceratina cyanea*). For this species, cut stems should not be removed, but laid on the ground (Edwards, 1996). Artificially removing surface vegetation, or digging scrapes to create nesting habitat for mining bees has proved successful in two Surrey localities (Edwards, reported in Dicks *et al.*, 2010), and at a degraded heathland site, in Oxfordshire, 80 species of bees and wasps colonised shallow bays cut into the sand over the following four years (Gregory & Wright, reported in Dicks *et al.*, 2010).

Fig. 6.37 A restored area of previously neglected heathland where populations of several specialist heath bees quickly recovered.

Mineral extraction sites

Sand and gravel quarries have proliferated alongside rapid urbanisation, but they provide an opportunity to compensate for habitat lost to building and industrial development (Fig. 6.38). Unfortunately most quarries are subject to infilling with municipal waste and are capped with soil for restoration of agriculture as soon as extraction ceases. Flooded pits are sometimes given over for angling, and, in a few cases, maintained as nature reserves. In either case, their margins and hinterlands can be readily colonised by bees, and some compare favourably with heather heaths. Particularly useful are south-facing steep slopes or vertical cliffs that characterise most workings. As with heathland, however, active management, preferably rotational, is required to retain areas of bare ground and ruderal herbaceous flowers, as well as patches of scrub, that are important for bees. Again, localised disturbance of the sandy soil is a useful management device in these habitats, though different considerations apply to agricultural set-aside (Corbet, 1995).

Calcareous grassland

At their best, south-facing, flowery downland slopes harbour a rich assemblage of solitary bees (Figs. 6.39–6.41). Hot microclimate, areas of bare ground and well-drained soil all provide favourable conditions for ground-nesting bees, and patches of scrub offer shelter and nesting sites for stem-nesting species. Wide rides and clear-fell in the forests of the East Anglian Brecks also sport complex mixtures of chalk and acidic grassland flora and suitable nesting habitat.

Fig. 6.38 A former mineral extraction site, with areas of bare, sandy ground suitable for mining bees.

Fig. 6.39 South-facing slopes of the North Downs. Prime bee habitat.

Fig. 6.40 Flower-rich grassland in the Suffolk Brecks

Fig. 6.41 Open ride with snail-shell nests of *Osmia bicolor* and *Osmia spinulosa*, Suffolk Brecks.

The scabious-specialists *Andrena hattorfiana* and *Andrena marginata* fly in such localities, as does *Melitta haemorrhoidalis*, a bee associated with wild and cultivated bellflowers. All three of the *Osmia* species which make their nests in snail shells can be found on chalk grassland and open woodland rides on chalk, but *Osmia spinulosa* and *Osmia bicolor* are mainly southern English species, while *Osmia aurulenta* also inhabits sites along the west coast of Wales, England and southern Scotland and in scattered colonies along the eastern coast of Ireland. It can also be found in some inland grasslands.

Former downland has been extensively ploughed and integrated into intensive arable cultivation, often with generous public subsidy, and much that remains has been allowed to develop into dense scrub and secondary woodland. Restoration to flower-rich grassland after abandonment is difficult (Baldock, 2008), hence the importance of maintaining suitable rotational management of remaining fragments.

Dry acid grassland

This is a much-threatened UK Biodiversity Action Plan priority habitat. Fine examples include Richmond Park and Wimbledon Common, to the south of London, and the grasslands of Epping Forest and the Thames Terraces to the east of London. Rapidly draining sandy soils, hot microclimates and areas of bare ground provide excellent bee nesting habitat, and the assemblages to be found here overlap with those of lowland heather heaths, though without the heather specialists. Both *Andrena cineraria* and *Andrena fulva*, two of our most distinctive mining bees, often abound in the open grassland, and this is also where their cuckoos *Nomada lathburiana* and *Nomada panzeri* will frequently be found.

Woodland

The value of woodland as habitat for bees varies greatly, depending on the underlying soil, extent, aspect, history and management (Figs. 6.42–6.44). Baldock (2008) describes the Surrey woodlands on the weald clay as rich in solitary bees, especially where there is Forestry Commission management with wide flowery rides and areas of clear fell. Large areas of conifer plantation on sand and chalk soils on the Brecks of East Anglia, also managed by the Forestry Commission, are especially rich in bees and wasps. Several species of *Andrena*, such as *Andrena wilkella*, *Andrena helvola*, *Andrena synadelpha*, *Andrena fucata* and *Andrena clarkella*, commonly inhabit open woodland, as does the mainly southern *Lasioglossum zonulum*.

Fig. 6.42 Coppice woodland management maintains open habitat for ground flora and invertebrates.

Fig. 6.43 A compacted south-facing earth bank in deciduous woodland, with nesting *Andrena clarkella*, *Andrena trimmerana*, *Andrena minutula* and others.

Fig. 6.44 Root-plate with nest burrows of several *Andrena* species.

Wetland

Except where wet areas are surrounded by bee-rich hinterlands, such as dry heath or past mineral workings, wetlands are not expected to support a large diversity of bee species. However, they are important for two localised species. *Macropis europaea* has a strong association with the edges of ditches or ponds in virtue of its dependence on yellow loosestrife, which grows in such places (Fig. 6.45). The other species is the yellow-faced bee, *Hylaeus pectoralis*, which nests in the disused galls of the fly *Lipara lucens* in the stems of common reed (Fig. 6.46).

Urban and suburban habitats

In the wake of habitat loss due to agricultural intensification, the attention of conservationists has turned to a range of habitats in urban areas, many of them humanly created, and often only incidentally of value to wildlife.

Abandoned industrial sites. So-called brown-field sites are frequently of high conservation importance, and especially rich in bees, wasps and ants (Fig. 6.47). Where the substrate is well-drained and nutrient-poor, patches of ruderal vegetation form, amongst areas of bare ground or pot-holed tarmac, irregular hollows and mounds, banks and old walls. These sites heat up quickly in the sun, and offer numerous places for bees to bask, nest, and forage. Where the soil allows, patches of scrub such as *Buddleja* or bramble develop, alongside numerous garden-escapes, while local bike-scramblers provide compacted paths as well as slowing down succession to scrub. Such places are anathema to the tidy-minded, and an irresistible temptation to the planner and developer. However, progress is being

Fig. 6.46 Habitat of *Hylaeus pectoralis*.

Fig. 6.45 Wetland habitat for *Macropis europaea* in east London.

Fig. 6.47 A Thames-side brown-field site, now a national nature reserve and home to an outstanding assemblage of bees and other invertebrates.

made in widening recognition of the wonderful reservoirs for wildlife that they can be. Organisations such as Buglife, as well as creative nature-writers, calling our attention to 'edgelands', have succeeded in saving at least some of our richest brown-field sites from development and ensuring their protection as reserves. Much then depends on subsequent management, but what should not be forgotten is the role of the informal recreation, often disapproved of, that served to 'manage' them in the past. Still, however, the abstract dichotomy of 'green-field good and brown-field bad' dominates much public thinking and official policy on development priorities. It is important that local natural history societies and Wildlife Trusts continue to emphasise the conservation and amenity value of many brown-field sites, and back up their campaigns with survey evidence.

Informal green spaces, churchyards, cemeteries, old orchards, parks and gardens. The long-established Local Authority method of management of green spaces is frequent motorised grass-mowing of the larger expanses of amenity grassland, combined with maintenance of formal beds of ornamental flowers. Even with such high levels of wildlife-hostile management, public parks have still retained populations of some of the more common, generalist bees and other insects. However, more ecologically enlightened ideas are slowly encroaching. Some LAs have found they can make considerable financial savings by replacing formal beds with specially planted wild-flower meadows, by leaving odd corners and edges unmown, and not removing dead wood after tree maintenance. Where grounds managers are open to such ideas, local naturalists

Fig. 6.48 A churchyard with abundant spring forage for bees..

may be able to influence further measures, such as creation of south-facing banks, leaving old walls un-pointed, siting bee hotels and planting bee-friendly flowers (Figs. 6.48–6.50). However, such changes as these can be alarming to members of the public who are used to traditional short-grass monocultures, and it must be remembered that public green spaces are not the monopoly of one interest group. Where there are voluntary 'partner' organisations involving interested members of the public, these provide a good context in which ideas for management change can be discussed before they are implemented. Such changes should, of course, be clearly explained and public feedback welcomed.

Those of us fortunate to have private gardens can manage to attract bees without such political challenges (though there can be important domestic negotiations to be had!). A small, town-centre garden may be quite capable of sustaining 30 or 40 species, and a larger, suburban garden in an area rich in bees may attract 100 or more (Fig. 6.51). As with other sites where conservation management is proposed, it is important to find out which species are already there. Until that is known, it is best to avoid radical change, especially if that might involve losing nest sites or flowers favoured by oligolectic species that might be present. In general, a garden favourable to solitary bees will have a range of flower-types that provide nectar and pollen from early spring through to late summer. Patches of scrub, hedges or trees, especially if composed of sallows or wllows (*Salix* species), fruit trees, laurel, flowering

Fig. 6.49 Nesting habitat for the scarce *Heriades truncorum* in a former landfill site, now country park.

Fig. 6.50 Even a formal park can provide valuable habitat for bees.

currant, rosemary, *Mahonia,* bramble, heathers, *Buddleja* and *Ceanothus,* can make an important contribution to continuity of forage sources. Roses, especially less horti-culturally modified varieties, provide nesting materials for leaf-cutters as well as floral resources.

Appropriate herbaceous plants may depend on local soils and geography, as well as the predations of slugs and snails, but several studies have highlighted a range of pollinator-friendly garden flowers (e.g. Comba *et al.,* 1999; Corbet *et al.,* 2001). Studies carried out in Sheffield by Smith *et al.* (2006a,b) examined the effects of wildlife gardening. Interestingly, both abundance and diversity of solitary bees was positively correlated with diversity of plant species, but negatively related to other aspects of wild-life-friendly gardening, notably provision for birds. Kirk & Howes (2012) is a valuable and comprehensive guide to bee-friendly flowers, with suggestions for short- and long-tongued guilds, and for the different seasons. Within their geographical range, it is also worth planting (or retaining) flowers for pollen specialists, such as bellflowers for *Melitta haemorrhoidalis,* white bryony for *Andrena florea* or ivy for *Colletes hederae.* Lamb's ears (*Stachys byzantina*) or yarrow (*Achillea*) afford the opportunity to watch female *Anthidium manicatum* clipping 'wool' from their leaves.

In common with other sorts of habitat, nest sites are as important as are floral resources. The nesting places of existing garden visitors should first be found, and where possible these should be protected. In the case of old walls or dilapidated cement rendering that is unavoidably repaired,

Fig. 6.51 This tiny town-centre garden provides habitat for as many as 30 bee species.

substitute nest-sites can be experimentally provided. Another possibility might be to deliberately leave holes in newly pointed walls, and monitor the results. A variety of commercial bee hotels is now on the market. These are often very successful in attracting bees – especially such species as the red mason bee, *Osmia bicornis* (formerly *Osmia rufa*) and some of the leaf-cutters. Dicks *et al.* (2010) review numerous experiments with artificial nests and discuss some successful designs. Placing an old tree trunk or beetle-bored fence-post in a sunny corner of the garden will often attract a range of other species, as will the provision of south-facing earthen or sandy banks.

The wider urban scene. The desirability of managing for conservation on a landscape scale does not apply solely to the countryside. Some species, such as *Anthophora plumipes,* are capable of foraging over several kilometres, and a local population may, likewise, range over a considerable segment of an urban landscape. A single garden managed for bees contributes a valuable 'oasis', but so much the better if adjacent gardens can be brought in to create a connected pattern of bee-friendly habitat. Conversations with neighbours, and even getting the support of neighbourhood associations, can result in quite large areas of private garden habitat functioning effectively as nature reserves.

But it is possible to work on a still more ambitious scale. Many Local Authorities (often with dubious justification) claim that their spatial planning attempts to apply the principles of Ebenezer Howard's vision of garden cities. There is also evidence-based interest in the public health benefits of informal green space and open-air exercise. Individuals and conservation organisations can make use of these arguments in consultations over planning policy to make the case for green spaces such as river courses, cemeteries, public parks and gardens, informal green spaces, local nature reserves, archaeological sites and even sports grounds to be mapped and conceptualised not as so many individual assets, but as a connected network that can be managed for both public amenity and wildlife. For them to be connected implies consolidating existing footpaths, bridleways, cycle tracks and permissive paths and adding new links to make movement between open spaces feasible for both wildlife and humans. A green infrastructure plan of this sort can offer solutions to a range of urban problems, as well as greatly enhancing local biodiversity.

7 Approaches to practical work

7.1 Introduction

In several places through this book I have mentioned limitations to current knowledge. Widespread alarm about the decline of many bumblebee species and well-known threats to domesticated honeybees has given rise to great advances in knowledge of them, partly through academic research and partly through contributions made by amateur observers. However, the shift of attention to our solitary bees has come much later. Now there is a degree of public investment in academic research into their role as pollinators, we are experiencing a rapid growth in our understanding of the complex, fascinating lives of this diverse group of insects. Amateur observers, too, are now making significant contributions, especially in mapping distributions, noting extensions and contractions of range in some species and charting the arrival of new ones.

In many ways this growth in core knowledge simply serves to expand the periphery of as-yet unanswered questions. Some of these questions may be for technical or economic reasons beyond the scope of most amateurs, but many others require little more than time, patience, a notebook and access to a park or garden. Persistent observation and recording of behaviour 'in the field' remains a crucial activity for obtaining insights into much that is currently unknown. If observation can be supplemented by stills photography and even video recording, then so much the better.

Once you have acquired some basic identification skills, then even more opportunities open up. Bees can be captured for more confident identification than is otherwise possible. Any observational discoveries gain greatly in significance if the species of their subjects is known with certainty, but also once captured the bees are available for closer study and measurement. Even basic experiments can be carried out by marking and releasing or manipulating bee behaviour.

For many observational studies, as well as more sophisticated experimental work, examination of dead specimens is unavoidable. This is something of a dilemma, as the emphasis in this book has been on appreciation of the lives and activities of bees, and the importance of conservation. Chapter 2, and its associated illustrations, together with Graham Collins's keys to genera (Chapter 8) have been designed to take you a long way towards bee identification without the need to kill specimens. A substantial minority (perhaps 40 to 50 species) of bees that occur in the British Isles can be identified from their general appearance as captured on a good photograph, or by examination of captive living insects using a x10 hand lens. Since many of these species are widespread and familiar, they can readily be used for observational and even some experimental studies without killing specimens.

However many readers of this book may wish to take their studies further, learning to identify and provide detailed studies of the many less distinctive and less well-known species. It is beyond the scope of this short book to give detailed advice on this, but if it is deemed justifiable to develop the ability to identify bees down to species level, a high-quality binocular microscope and other equipment will be needed. Bees can be killed by placing them in a freezer for a period, a method that has some disadvantages, but dispenses with the need to handle chemical poisons. When dead they can be set by piercing the thorax with a fine pin, and body-parts arranged so as to show features necessary for identification. Opening the mandibles, and, in the case of the males of difficult genera such as *Andrena* and *Lasioglossum*, extruding genital capsules with a fine needle, may be necessary (Fig. 7.1). Various types and sizes of pin can be obtained from a number of entomological suppliers, many of which have equipment on sale at the Annual Exhibition of the Amateur Entomologists' Society in the UK (amentsoc.org). It is essential that a label is attached to the pin giving the date and location of the capture, preferably cross-referenced to notebook details of

genital capsule

a complex structure including the penis and associated organs that is contained in a cavity at the rear of the abdomen of male bees, and extruded to enable mating and insemination of females

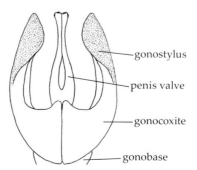

Fig. 7.1 The genital capsule of a male *Andrena* species.

behaviour and photographs if available. A small reference collection can be made, and kept in an airtight store-box lined with cork or polythene foam. As mentioned elsewhere, we now have an excellent comprehensive and up-to-date guide to the bees of the British Isles (Falk & Lewington 2015), and another is in preparation. In addition there are other publications that illustrate and describe many species (see Chapter 2) and websites such as that of the Bees, Wasps and Ants Recording Society. Local natural history societies, Wildlife Trusts or organisations such as the RSPB will often have experts willing to help with identification, supply draft keys or provide a reference collection. Every effort should be made to learn to recognise bees in the field, so that the need to kill and collect specimens can be kept to a minimum.

7.2 Photography and video

For many species, one or more sharply focused photographs will enable reasonably confident identification. It is helpful to know in advance the sort of features that are used in identification. If these are visible when the insect is alive and behaving as usual in its habitat, then the ideal is to obtain an image taken from the right angle to show the features in question. These might, depending on the group concerned, include wing venation, facial features, patterns and colours of hair on the body (such as colours of the pollen-hairs on the underside of females in the Megachilidae), shapes and colour-patterns on the legs, and so on. To capture an image of wing venation it is sometimes possible to photograph an insect on the move with its wings spread (and, preferably, with a pale background). However, this is best done by catching the insect and constraining it in a narrow plastic or glass tube. It can be slowed down by carefully pressing on it with a wad of tissue introduced into the tube. Alternatively, the tube containing the insect can be placed in a refrigerator for an hour or so. Its wings can then be spread out and photographed while it warms up and recovers before being allowed to fly off.

First, of course, the bee must be caught. There are several approaches to this. The method I find most useful is to observe the target bee as it forages from a patch of flowers. Ideally, one stands by the flowers and waits for the bees to arrive, but if it is necessary to approach them, this should be done very slowly, and without casting a shadow on them. Then an open tube can be carefully placed over the bee as the lid is brought up from below. Often this involves catching the bee and the flower together and the bee sometimes just continues feeding. If the intention is to use photographs of the captive bee for identification, then it is desirable (but often not possible) to take photographs of the bee as it forages naturally before capturing it, so that you have two sources of information.

I hesitate to advise on photographic equipment, as technical advances are now taking place so rapidly. I use a fairly expensive SLR camera with a fixed focus macro lens and mounted flash units. This allows me to get close-up shots of all but the smallest bees while still working at sufficient distance that they are not frightened away. However, especially for identification shots of captive insects, an inexpensive camera with a macro setting, or a mobile phone will produce excellent results.

However, both photography and video can be used for many purposes other than identification. Still shots can often be very revealing of behaviour, showing insects collecting nectar or pollen from different flower species, interacting in various ways during courtship and mating, constructing their nests and so on. However, producing video with a video camera or a stills camera with video options can be even more instructive, capturing whole sequences of behaviour with a degree of detail not perceptible by unaided observation.

7.3 Observational studies

Much can be discovered simply by persistent observation and note-taking. Many aspects of the relationship between bees and flowers can be studied in this way. An inventory can be made of the variety of plants currently in flower in a given local patch (a garden, public park, nature reserve, brown-field site or whatever). Observing stands of each flower for a standard period of time, the number of visits by bees of different species and/or other insect groups can be recorded. Do some flowers attract a wider range of insects than others? Are there noticeable differences between the groups of insects that are observed to visit flowers of different shapes, sizes, colours and so on? Do particular species of bee show foraging preferences for particular groups of flowers? Are some bees more choosy about which flowers to visit than others?

With more experience, it should be possible to judge whether female bees are foraging for nectar or pollen. Does their pattern of foraging differ according to which reward they are seeking? Here it may be interesting to film the sequence of behaviours involved in collecting, transferring and carrying pollen from one flower to the next. Does the bee show flower constancy? How frequently does it stop to brush pollen from its body fur onto its scopae? As this might involve following the same bee for a relatively prolonged period it may be helpful to find a way of identifying it. Older bees often have signs of damage, especially to their wings, or loss of hair from the thorax and this can be used to recognise individuals. A more reliable method is to capture the bee in a tube of the sort used by beekeepers to mark queen bees. This has a plunger at one end, and netting across the other. A captive bee can be pressed against the netting, and a tiny dab of paint, erasing fluid or even nail-varnish applied to the back of its thorax.

Once you have a way of individuating bees it is possible to devise more advanced observational studies. Chapter 3 of this book reports on the studies carried out by Graham Stone of the daily routines of *Anthophora plumipes* males and females. The emphasis in his studies was on the effects of temperature changes through the day as well as the impact of sexual harassment by males on the foraging strategies of the females. These studies could be used as models for investigating foraging behaviour of other species. In Chapter 5, it was suggested that one use of the notion of a competition box (Corbet *et al.*, 1995) might be to investigate the way assemblages of flower-visitors change throughout the day. Variables such as the sizes of the insects, their tongue lengths, the range of flower-types in the site under study, the standing crop of nectar and temperature changes through the day are likely to affect which species are foraging at each time of day. Are such shifts of species-composition through the day actually observed? Is there evidence of direct competitive interaction between individuals of the same or different species? An interesting and related project might be to note the threshold temperatures at which different species become active early in the day, and cease activity at the end of the day. Is the expected link between size and ability to forage under adverse conditions borne out? If not, what other factors might be involved? Is there a correlation between the temperature threshold for foraging on a daily basis and the seasonal flight-period of the bee (phenology)? Again, is there a relationship between the temperature threshold at which individuals of a species can be active and their geographical distribution?

Academic research on bees has tended to be focused on foraging activity, with relatively less known about other aspects of bee lives. The widespread use of bee hotels in gardens and some public parks provides excellent opportunities for observing nesting behaviour among stem and crevice nesting species. One valuable exercise might be to design different models (or compare different ones already on the market). Using standard locations (in terms of height, aspect and exposure), different design features can be compared for their attractiveness to different species. How does the width of the bore of canes or tubes affect if bees utilise them, and which species do so? How do holes bored into blocks of wood compare with tubes or lengths of cane? Is a log or fence-post already occupied by wood-boring insects of various kinds and translocated to the garden more successful in attracting nests than an artificial bee box?

Once nesting has become established, whatever the means used, other questions can be addressed. If the above-mentioned technique for marking individual bees is employed, then, for example, the rate at which each cell is constructed and fully provisioned can be studied for a range of weather conditions and nesting species. Also, as in some species mate-location takes place around the natal nests of the females, mate-location, female choice, courtship and mating behaviour can be studied, and recorded using still photos or video. Artificial nest-sites such as these also offer valuable opportunities to observe the behaviour of the cleptoparasitic species. In some cases, even the range of hosts used by a species of *Nomada* or *Sphecodes* is not yet fully known, as are many details of how nests are discovered and approached by nest-parasites. Neither are there many reports of defensive responses, if any, on the part of the hosts.

Unfortunately since so much of bee behaviour takes place within the nest it is normally out of sight of human observers. However, there are artificial nest designs that make use of glass tubes which do allow visual observation of nest construction and other activities. Baldock (2008) describes and illustrates one such design (developed by Kim Taylor). Baldock has succeeded in attracting as many as 121 species of bee to nest in his own garden using such methods.

Finally, once the beginner has learned to identify even a few of the most distinctive bees a useful contribution can be made to surveying local sites (including gardens, which, of course, are not normally accessible to anyone but the householders), which may not have been studied. Lists of species present can make a valuable contribution to the development of management plans, and can also be used to make a case for protection of a site from adverse developments. Information on the presence of each species in a locality can be passed on to the appropriate land-manager (especially if that is a local wildlife organisation, local authority or sympathetic private landowner) for this purpose. Also, each county should have at least one organisation that systematically collects and assesses wildlife records, and a county recorder for each group of organisms. County-wide distribution maps are of vital importance in assessing the conservation status of each species, and so determining priorities for conservation action. Relatively little expertise is needed to make a valuable contribution to collective knowledge of the status of bees, in particular, in your area. There are various preferred ways of submitting records, but crucially important are: species name, locality, exact map reference and date. Also, and especially if it is an unusual discovery, or relates to a species that is difficult to identify with certainty, the name of the person who confirmed the identification should be given. If there is a good photograph (or, with the above provisos, a specimen) this will help to verify the record.

Above all, it is important to keep a detailed record of your observations and ensure that any of more general interest are made public – for example by writing a note for publication in a wildlife magazine. This might be a newsletter of a local wildlife or natural history society, or even in one of the national entomological journals. Even if an article is not accepted, editors will generally give advice on how to prepare an article for publication.

8 Keys to the genera of bees of the British Isles

by Graham A. Collins

8.1 Introduction

There are around 250 species of bee occurring in Britain. They are an interesting group to study because of their biology and ecology and are attractive insects in themselves. Identification of bees needs to start with identification to generic level; once more experience has been gained it is usually obvious what genus is involved just from the look of the insect.

The keys presented here should enable almost all bee specimens to be assigned to a genus. It is designed for the beginner and complex technical terms are kept to a minimum. A glossary is provided and most features used in the keys are illustrated in the relevant couplet. Ease of use is paramount and long, complex couplets allowing for every eventuality have been avoided – so from time to time a particular specimen might prove difficult or impossible to key out, especially when only a few genera have been encountered. Put it to one side and try again when more material has been accumulated and you have more experience. The keys will only work with examples of the genera from the British Isles and other works will need to be consulted for European material.

8.2 Checklist – is it a bee?

Many insects mimic bees while at the same time a number of bees, mostly cleptoparasitic species, are distinctly wasp-like. Before working through the keys it is sensible to check that your insect is a bee!

- Two pairs of membranous wings; the forewings with 9–10 enclosed cells, the hindwing with a row of hooks on its front edge which connect it to the forewing in flight.

- Mandibles present, between which a tongue is usually visible.

- Antennae with 12–13 segments (be careful, the second segment can be very short and partially hidden within the first; however, the third is usually long and distinctly conical, differing from the following segments).

- Distinct constriction between thorax and abdomen ('wasp-waist'), not easy to see in very hairy species.

- Plumose hairs; branched hairs adapted to carry pollen, these are usually obvious in non-parasitic species but in parasitic bees only a few remain, particularly on the propodeum and below the thorax. Simple hairs may also be present.

- First segment of hind tarsus usually flatter and wider than following segments.

8.3 How to use the keys

Having killed the bee, preferably in the fumes of ethyl acetate but, as an alternative, 24 hours in a domestic freezer will suffice, it should be mounted. Continental entomological pins are recommended to facilitate handling and to allow multiple labelling. The insect should be positioned about one-third from the head so that enough room is left to manipulate the pin without damaging the bee. Appendages should be moved away from the body, the mandibles opened if possible (don't force them, the jaw muscles are much stronger than the neck muscles and loss of the head is almost inevitable). In addition, if the bee is a male (see Key A), the genital capsule should be extracted from the gaster using a pin. Ideally it will be fully visible but still attached to the body. If it is necessary to remove it completely, it can be mounted on a piece of card attached to the same pin as the bee. All specimens should be labelled with collecting data (at least site, vice-county, full grid reference, full date and collector), and, once identified, a determination label with the name of the species, the determiner and the date determined.

vice-counties
a geographical division of the British Isles used for biological recording, based on the ancient counties, but with many counties divided into smaller units to make them more uniform in size

Separate keys are given to males and females. In many cases males and females of the same species are distinctly more different than the same sexes of closely related species. Identifying a bee to genus is only the first step in the process, identification to species is the ultimate aim and most published keys treat the sexes separately. It is thus sensible to get used to recognising males and females from the start. Failure to do so correctly will cause problems as different characters are used in the two keys.

Check the insect agrees with the characters listed above and move on to Key A. Each key consists of couplets which list alternative character states. In these keys they are subdivided into clauses, with, for example, the alternatives being a or aa, b or bb. Where possible the description is accompanied by a figure, positioned above or to the left of the couplet, that illustrates the position and state of each character, but a separate glossary is also given. Clauses are presented in order such that easy to see, definitive characters are used before more variable and comparative ones. Tongue characters, used in a number of other keys, are only used where absolutely necessary as the tongue may well be hidden. Characters of wing venation are used widely. They are generally very constant, but occasionally particular veins can be wholly or partially absent, although often traces remain, especially at the junction with other veins; it is best to check both wings. Starting at couplet 1, read each clause and its alternative before making a decision – each half of the couplet will then lead on to either another couplet or the answer. If there appears to be a conflict between the two halves of a couplet you may have gone wrong earlier. You have two options. Either follow each lead and see if the situation resolves, or go back to the previous couplet and check it again. The number of the previous couplet is given in brackets next to the current couplet number.

8.4 Glossary

Arolium
A small pad which projects between the two claws of the foot (Fig. 8.1).

Axillae
Small triangular plates on either side of the scutellum, usually inconspicuous but in some genera enlarged and projecting backwards as teeth (Fig. 8.4).

Claval lobe
see Jugal lobe (Fig. 8.6).

Cleptoparasite
A species that steals the nest of another. The host species creates and stocks a nest and lays her egg in it. The parasite opens the nest and lays her own egg which eats the host's resources. The host egg is destroyed by the parasite female or her larva.

Clypeus
The central plate of the face (Fig. 8.3).

Cubital cell
see Jugal lobe (Fig. 8.6).

Gaster
The equivalent of the abdomen in other insects; in the apocritan Hymenoptera the first segment of the abdomen is integral to the thorax and the 'wasp waist' actually occurs between the first and second segments (cf. propodeum). Also known as the metasoma (i.e. Fig. 8.2).

Integument
A hard external covering

Jugal lobe
The lowermost longitudinal vein of the hindwing, if followed to the wing margin, reaches a point where the wing border is excised. The lobe thus formed behind

the vein is the claval lobe. If the margin is followed back to the wing base a second incision may be present marking off the jugal lobe. The presence and length of the jugal lobe relative to the claval lobe and the cubital cell is an important character in the keys; in comparison, both are measured from the wing base (Fig. 8.6).

Labrum The plate at the base of the tongue, just below the clypeus (Fig. 8.3).

Marginal area The apical part of each tergite, usually one third to one half its total length, often with different vestiture (hairs) or punctation (punctures) to the basal part (Fig. 8.2).

Mesonotum The top of the central part of the thorax, between the wing bases (Fig. 8.4).

Parapsidal line A modified area on the mesonotum, usually linear, lying level with the wing-base and half way between base of wing and mid-line (Fig. 8.4).

Propodeum The back of the 'thorax' (or mesosoma) is actually the first segment of the abdomen (Fig. 8.4).

Pygidium A modified, usually triangular, area on the last tergite of the gaster, usually in females where it is used to remove material when nest-building (Fig. 8.2).

Scape The elongate, basal segment of the antenna.

Scopa The pollen collecting apparatus of the female, consisting of dense and long hairs. In some genera it is present below the gaster, in others on the hind leg, particularly the tibia and first tarsal segment. It is lacking in parasitic species.

Scutellum The part of the thorax behind the mesonotum (Fig. 8.4).

Sternites The plates that make up the lower surface of the gaster.

Tegula A small plate covering the base of the wing (Fig. 8.4).

Tergites The plates that make up the upper surface of the gaster.

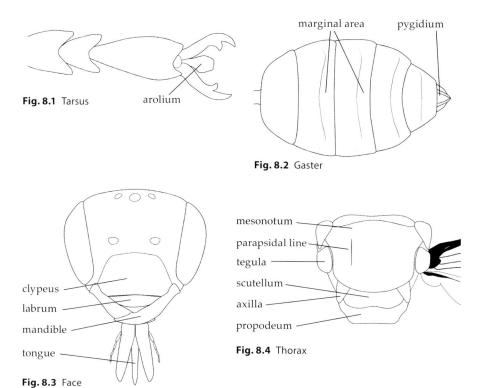

Fig. 8.1 Tarsus

arolium

marginal area

pygidium

Fig. 8.2 Gaster

clypeus

labrum

mandible

tongue

Fig. 8.3 Face

mesonotum

parapsidal line

tegula

scutellum

axilla

propodeum

Fig. 8.4 Thorax

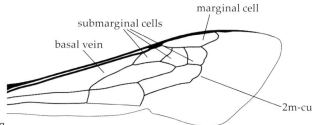

marginal cell

submarginal cells

basal vein

2m-cu

Fig. 8.5 Forewing

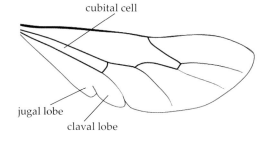

cubital cell

Fig. 8.6 Hindwing

jugal lobe

claval lobe

[1] The antenna comprises: a scape (the elongate basal segment), a pedicel (often very small), and a number of flagellar segments. Sometimes the pedicel is so small that it is lost within the scape and an incorrect segmental count may result. An alternative way of counting is that the first flagellar segment (third segment of the antenna) is often longer and more conical than the following segments. Start with this one and count to the end, then add 2. Sometimes it helps to count from the tip towards the base.

[2] Very rarely, aberrant individuals of species with three submarginal cells appear to have only two through the loss or reduction of a cross-vein. Often this missing cross-vein can be seen as a vestige, or as short appendixes, at its junction with the other veins. If entirely absent you will be forced to follow the wrong half of the couplet and the key will fail at a later point or the bee will not match the criteria in the generic description – in this case you should try the option for three submarginal cells.

b

Key A

1	a	Antennae with 12 segments[1]
	b	Gaster with six visible tergites
	c	Sting present, which may protrude after death
	d	Scopa, or pollen collecting apparatus, often present, either on the hind legs or below the gaster (absent in the parasitic, usually wasp-like, species)

Females – Key B

–	aa	Antenna with 13 segments[1]
	bb	Gaster with seven visible tergites (but in some genera, the apical ones folded beneath the end of gaster)
	cc	Complex internal genitalia present in the form of a capsule (which should be hooked out with a pin to facilitate identification in certain genera)
	dd	Never with scopa (although incidental collection of pollen may occur through foraging for nectar)

Males – Key C

Key B – Females

a   aa

1	a	Forewing with two submarginal cells[2]	**2**
–	aa	Forewing with three submarginal cells	**15**
2 (1)	a	Surface of eyes with long dense hairs	
	b	Gaster strongly narrowing to pointed apex	

Coelioxys

Medium-sized to large bees (9–16 mm); gaster rather shining and with pale bands or wedge-shaped spots formed of flattened hairs; scutellum with rearward-pointing projections either side; no scopa. Cleptoparasitic on *Megachile* and *Anthophora*.

| – | aa | Surface of eyes bare | |
| | bb | Gaster more-or-less parallel sided or oval, rounded at apex | **3** |

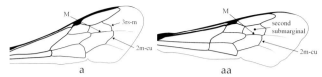

3 (2) a   Vein 2m-cu meets M beyond point where 3rs-m
            does   **4**

—   aa   Vein 2m-cu meets M opposite second
            submarginal cell   **5**

4 (3) a   Gaster black or at most with cream-coloured
            spots or bands, face and legs without colour

       b   No scopa present

       c   Arolium present between tarsal claws ***Stelis***

            Small to medium-sized species (5–11 mm); black,
            sometimes with pale markings on gaster; rather
            shining, heavily-armoured species. Cleptoparasitic
            on *Osmia, Hoplitis, Anthidium* and *Heriades.*

—   aa   Gaster with bright yellow spots, similar colour
            present on face and legs

       bb   Dense golden-yellow scopa on underside of
            gaster

       cc   No arolium between tarsal claws ***Anthidium***

            One British species – *manicatum.* Large (11–15
            mm); black with yellow spots on gaster, tibiae,
            sides of mesonotum, tegulae, top of head, face and
            mandibles.

5 (3) a   Arolium present between tarsal claws   **6**

—   aa   Arolium absent   ***Megachile***

            Medium-sized to large species (9–18 mm); head
            large, mandibles triangular with broad cutting edge
            carrying several teeth; scopa present on underside
            of gaster. 'Leaf-cutter bees'.

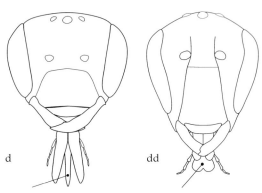

d   dd

6 (5)   a   Scopa (pollen collecting hairs) present, either on hind leg or underside of gaster

        b   Face black beneath any hairs

        c   Legs black, without yellow markings

        d   Tongue pointed at apex, fairly short to long **7**

–   aa No obvious scopa present

        bb EITHER yellow markings present on face OR face black with projecting lobes at lower corners of clypeus and a bulge below antennal bases forming a triangular concavity

        cc Legs almost hairless and with clear yellow markings

        dd Tongue short and bilobed at apex   ***Hylaeus***
            Small to medium-sized species (4–8 mm); very sparsely-haired bees; black with yellow on legs and (usually) face.

7 (6)   a   Scopa on hind legs   **8**

–   aa Scopa on underside of gaster   **12**

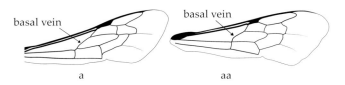

basal vein   basal vein

a   aa

8 (7)   a   Basal vein (second section of M) almost straight   **9**

–   aa Basal vein (second section of M) fairly strongly arched   **10**

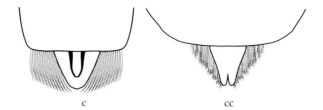

c   cc

9 (8)   a   Mesonotum very sparsely haired; surface
              brightly shining between the punctures

         b   Gaster sparsely haired, except for the apical
              tergites, without hint of pale bands

         c   Pygidium triangular with a blunt or rounded
              apex and raised median keel   ***Panurgus***

              Medium-sized species (7–10 mm); shiny black with
              pale, long haired scopa.

–   aa   Mesonotum densely haired with a mixture of
              black and golden-yellow hairs; surface matt
              between the punctures

         bb   Gaster with apical half of tergites 2–4 covered
              with dense, pale flattened hairs, contrasting
              with the sparser black hairs of the basal halves,
              thus appearing banded

         cc   Pygidium long triangular with a deeply
              notched apex, its surface flat   ***Dasypoda***

              One British species – *hirtipes*. Large (13–15 mm);
              banded bee with conspicuous scopa, the hairs of
              which are considerably longer then the thickness
              of the tibia.

10 (8)   a   Small species, not over 7 mm

         b   Tergite 4 of gaster with sparse, erect hairs only

         c   Third antennal segment very short, shorter
              than pedicel and similar in length to fourth
              segment   ***Dufourea***

              Two British species, both very rare or extinct.
              Resembling a small *Lasioglossum*, but with only two
              submarginal cells in the forewing. 5–7 mm.

–   aa   Medium-sized to large species, 9–15 mm

         bb   Tergite 4 of gaster with a band of pale, dense,
              flattened scale-like hairs

         cc   Third antennal segment elongate, much longer
              than pedicel and nearly twice as long as fourth
              segment   **11**

11 (10) a   Medium-sized species, not over 11 mm

          b   First tergite smooth and shining, with small widely-spaced punctures, almost hairless

          c   Marginal area of tergite 2 smooth and without punctures   ***Macropis***

               One British species – *europaea*. Shining black bee, with pale hair-bands on the apical tergites. Associated with yellow loosestrife.

–   aa   Large species, 13–15 mm

          bb   First tergite densely covered with deep punctures, from which arise long hairs

          cc   Marginal area of tergite 2 rather densely punctate and with surface dulled   ***Eucera***

               Large, dull species, with pale bands on apical tergites.

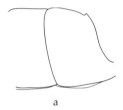 

a   aa

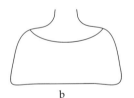

b

12 (7) a   First tergite of gaster with a strongly raised sharp ridge at front, separating dorsal surface from anterior face

         b   Viewed from above, this ridge distinctly concave   ***Heriades***

             Two British species - *adunca* and *truncorum*. Medium-sized (7–8 mm), rather slender bee; body shining with dense, deep punctures; clypeus with paired median apical tubercles.

–   aa   First tergite of gaster more-or-less smoothly curved from dorsal surface to anterior face, without a distinct transverse keel

         bb   If change in surface texture between dorsal surface and anterior face gives the impression of a vague ridge, this ridge viewed from above almost straight   **13**

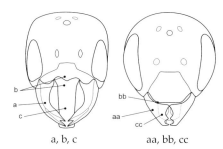

a, b, c   aa, bb, cc

13 (12) a Mandibles long and narrow, tapering towards apex

b At rest, upper edge of mandibles nowhere near clypeus, but leaving an opening through which the labrum is clearly visible

c Labrum very long, so that its tip is visible below apices of closed mandibles

d Thorax elongate, area behind scutellum horizontal, almost as long as vertical posterior face of propodeum ***Chelostoma***

Small to medium-sized species (5–11 mm), shining black with elongate abdomen.

– aa Mandibles shorter, length less than twice basal width, parallel sided or widening towards apex

bb At rest, upper edge of mandibles fitting closely against edge of clypeus

cc Labrum longer than wide but not projecting below apices of closed mandibles

dd Thorax short, falling away vertically immediately behind scutellum **14**

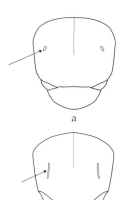

14 (13) a On mesonotum, parapsidal lines short, scarcely longer than wide, appearing as a raised, flattened area usually distinct from surrounding punctures (move specimen relative to light-source to create reflections) AND/OR

b Scopal hairs golden-reddish or black ***Osmia***

Medium-sized to large species (7–14 mm); many either with long, dense red hair or metallic integument. 'Mason bees'.

– aa On mesonotum, parapsidal lines linear, many times longer than wide, not always obvious AND

bb Scopal hairs white ***Hoplitis***

Medium-sized species (6–10 mm); sparsely haired with black integument.

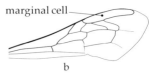

marginal cell

b

15 (1) a Surface of eyes with long dense hairs

b Marginal cell long and narrow ***Apis***

The honeybee.

– aa Surface of eyes bare

bb Marginal cell usually shorter and broader **16**

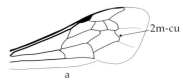

2m-cu

a

16 (15) a Vein 2m-cu strongly S-shaped, the lower end bulging outwards

b Tongue short and bilobed at apex

c Head, viewed from in front, rather triangular, the inner margins of the eyes converging ventrally ***Colletes***

Medium-sized to large bees (8–16 mm), most species with dense flattened hairs covering the marginal areas of the tergites, producing a banded effect; scopa on hind legs.

b, c

– aa Vein 2m-cu usually straight, at most slightly curved and then not bowed outward at lower end (illustration, next couplet)

bb Tongue variable in length but always pointed at apex

cc Head more rounded or oval, inner margins of eyes usually more parallel **17**

bb, cc

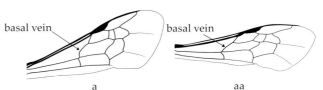

basal vein basal vein

a aa

17 (16) a Basal vein strongly arched with a distinct bend towards lower end, forming almost a right angle where it meets the longitudinal vein **18**

– aa Basal vein almost straight or slightly and evenly arched, the lower end meeting the longitudinal vein at an acute angle **21**

a

18 (17) a   Fifth tergite with a specialised hair patch, the rima, in the form of a central, longitudinal bare area surrounded on each side by dense, flattened hairs

     b   Scopa on hind legs

     c   Integument wholly black or metallic blue or green   **19**

–   aa   Fifth tergite without rima, either almost bare or with flattened hairs forming a complete apical band

     bb   No obvious scopa

     cc   Integument either metallic blue or with some tergites clear blood-red   **20**

19 (18) a   Gaster with bands or spots of whitish flattened hairs situated on the marginal areas of the tergites (glossary, Fig. 8.2), often extending beyond the apical margin and thus masking it

     b   Outer cross-veins similar in thickness and pigmentation to adjacent wing veins   *Halictus*

     Small to medium-sized species (6–11 mm); integument black or metallic bronzy green.

–   aa   If gaster with patches of whitish flattened hairs, these are situated basally, often originating below the apical margin of the preceding tergite which is thus obvious; in some species these hair patches absent or indistinct

     bb   Cross-veins towards wing tip usually finer and less obviously pigmented than adjacent longitudinal veins   *Lasioglossum*

     Small to medium-sized species (5–11 mm); integument black or metallic greenish to bluish.

20 (18) a   Integument obviously metallic bluish

     Ceratina

     One British species – *cyanea*. Fairly small (6–7 mm), shining metallic blue bee. Nests, and overwinters, in bramble stems. The 'small carpenter bee'.

–   aa   Integument black, with several tergites completely blood-red   *Sphecodes*

     Very small to medium-sized species (4–12 mm), gaster black with more-or-less extensive red belt, usually rather shining; heavily punctured head and thorax. Cleptoparasitic on species of *Lasioglossum*, *Halictus* and *Andrena*.

21 (17) a   Wings strongly purplish-iridescent

       b   Very large species, over 18 mm, with entirely black hairs on body and legs

Xylocopa

One species – *violacea* – a vagrant to Britain, but seen more often in recent years and could become established.

–   aa   Wings usually clear, at most smoky brownish

      bb   Often smaller, if as large as 18 mm then body usually with bands or spots of lighter coloured hairs   **22**

         a   aa

22 (21) a   Lower border of third submarginal cell (measured between intersection with its bordering cross-veins) distinctly longer than that of second submarginal cell   **23**

–   aa   Lower border of third submarginal cell more-or-less equal in length to that of second or even shorter   **25**

23 (22) a   Fairly distinct scopa on hind legs

       b   Integument dark or with reddish marks or bands, not metallic   **24**

–   aa   No evident scopa

      bb   Integument distinctly metallic bluish

Ceratina

One British species – *cyanea*. Fairly small (6–7 mm), shining metallic blue bee. Nests, and overwinters, in bramble stems. The 'small carpenter bee'.

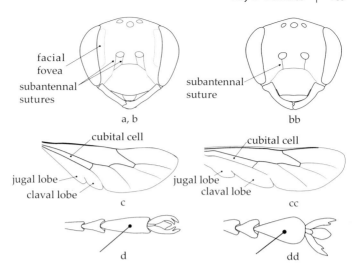

24 (23) a Facial foveae present – distinct depressions just inside inner border of eyes and lined with very short, dense hairs giving a velvety appearance

          b Two subantennal sutures present, showing as shining lines descending from antennal socket to meet upper border of clypeus, the shining suture defining the clypeus slightly thickened at these points

          c Jugal lobe of hindwing long, its length, measured from base of wing, distinctly more than half the length of claval lobe, usually reaching as far out as the vein closing the cubital cell (care needed – it may be folded under)

          d Last tarsal segment slender, at least three times as long as wide   ***Andrena***

            Small to large species (6–15 mm); a very large and diverse genus with species ranging from almost hairless to densely haired, shining to dull, some banded, and some with reddish markings on gaster.

   –   aa Facial foveae absent, the face inside eyes not depressed or lined with velvety hairs

         bb One subantennal suture present

         cc Jugal lobe of hindwing short, less than half length of claval lobe, and not nearly reaching vein closing the cubital cell

         dd Last tarsal segment broad, about twice as long as wide   ***Melitta***

            Medium-sized to large species (10–15 mm); integument dark; gaster with either whitish hair bands or an orange tail.

25 (22) a   One or more pairs of legs marked with or completely yellow or reddish-orange **26**

–   aa   All legs with integument entirely black **27**

26 (25) a   Gaster with paired pale spots formed from dense flattened hairs

   b   Axillae, on either side of scutellum, large and triangular, projecting backwards as a pair of teeth   ***Epeolus***

Medium-sized species (6–11 mm); cleptoparasites of *Colletes*.

b

bb

–   aa   Gaster without patches of dense flattened hairs, instead patterned by yellow, red or brownish spots or bands due to pigmentation of the integument

   bb   Axillae, on either side of scutellum, small and inconspicuous, not projecting backwards as teeth   ***Nomada***

Small to large species (4–15 mm); gaster patterned with red or yellow and shining (thus wasp-like), head and thorax often heavily punctured. Cleptoparasites of *Andrena*, *Lasioglossum*, *Melitta* and *Eucera*.

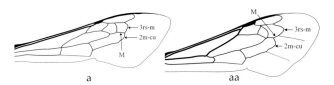

a   aa

27 (25) a   Veins 2m-cu and 3rs-m touching where they meet vein M

   b   Scopa present on hind leg   ***Anthophora***

Fairly to very large species (10–17 mm); usually rather hairy; gaster with hair bands in some species; eyes, in life, sometimes rather greenish.

–   aa   Vein 2m-cu meeting M clearly nearer base of wing than 3rs-m does

   bb   Scopa either entirely absent, or pollen-collecting apparatus present in the form of a corbiculum – the hind tibia flat and shiny and bordered by long hairs   **28**

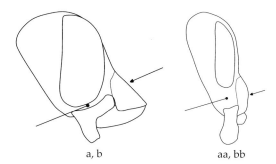

a, b   aa, bb

28 (27) a   Cheek (malar space) short; distance from lower margin of eye to mandibular base less than one fifth of width of mandible here

      b   Face strongly protruding, clypeus at about 45º to vertical

      c   Gaster mainly sparsely haired and shining, with very long hairs confined to base, tergites with paired lateral spots formed from flattened hairs   ***Melecta***

         Large to very large bees (12–16 mm), black with greyish-white hair; scutellum with paired apical prongs (difficult to see beneath hair). Cleptoparasites of *Anthophora*.

–   aa   Cheek (malar space) long, equal to or longer than width of mandibular base

      bb   Face almost flat, clypeus vertical

      cc   Gaster relatively densely long-haired, especially towards apex, never with rounded spots but often with complete bands of coloured hair   ***Bombus***

         Medium-sized to very large species (10–35 mm); bumblebees, with pollen-basket, and cuckoo bumblebees, without. (Includes the former genus *Psithyrus* now considered a subgenus of *Bombus*).

Key C – Males

a   aa

1   a   Forewing with two submarginal cells[3]   **2**

–   aa   Forewing with three submarginal cells   **15**

2 (1)   a   Antennae exceptionally long, at least as long as forewing   ***Eucera***

Large species (12–16 mm); rather hairy; face yellow.

–   aa   Antennae normal, rarely more than half as long as forewing   **3**

3 (2)   a   Face with yellow markings   **4**

–   aa   Face with integument black, although it may be covered with dense pale hairs   **6**

4 (3)   a   Gaster with paired, lateral, bright yellow spots

b   Tergites 6 and 7 of gaster with prominent, hook-like projections   ***Anthidium***

One British species – *manicatum*. Large (11–15 mm); black with yellow spots on gaster, tibiae, tegulae, top of head, face and mandibles.

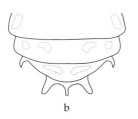

b

–   aa   Gaster completely black

bb   Apical tergites usually unadorned, at most with a median, down-curved projection on last tergite   **5**

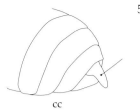

cc

5 (4)   a   Legs with yellow spots on at least tibiae or tarsi

        b   Legs slender, hind tibia oval in cross-section

        c   Apex of gaster simple   *Hylaeus*

            Small to medium-sized species (4–8 mm); very sparsely-haired bees; black with yellow on legs, face and sometimes parts of thorax; tongue short with bilobed apex.

   –   aa Legs completely black

        bb Legs robust, hind femur and tibia strongly swollen, the latter rather triangular in cross-section

        cc Last tergite of gaster with median, down-curved, tongue-like projection   *Macropis*

            One British species – *europaea*. Shining black bee, with pale hair-bands on the apical tergites and yellow face. Associated with yellow loosestrife.

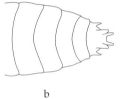

b

6 (3)   a   Surface of eyes with long dense hairs

        b   Tergite 6 (apparent apex) of gaster with three pairs of posteriorly projecting spines   *Coelioxys*

            Medium-sized to large species (9–14 mm); head and thorax densely and deeply punctured; gaster black and shiny, tapered towards apex, with bands of white, flattened hairs; scutellum with rearward-pointing hooks either side. Cleptoparasitic on *Megachile* and *Anthophora*.

   –   aa Surface of eyes bare

        bb Tergite 6 without three pairs of spines, at most with roughened rim   7

arolium
        a   aa

7 (6)   a   Arolium present between tarsal claws   8

   –   aa Arolium, between tarsal claws, absent   *Megachile*

            Medium-sized to large species (9–15 mm); mandible with broad cutting edge with 3–4 teeth; tongue long; some species with front tarsus modified, expanded and pale-coloured. 'Leaf-cutter bees'.

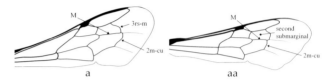

a   aa

8 (7)   a   Vein 2m-cu reaching vein M at or beyond the point that vein 3rs-m does   *Stelis*

Small to medium-sized species (5–10 mm); gaster black, black with narrow pale margins to tergites or black with pale lateral spots; rather shining, heavily-armoured species. Cleptoparasitic on *Osmia*, *Hoplitis*, *Anthidium* and *Heriades*.

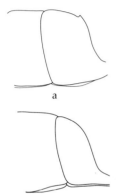

–   aa   Vein 2m-cu reaching vein M before 3rs-m does, thus entering second submarginal cell   **9**

9 (8)   a   First tergite of gaster with a strong transverse ridge at front, separating dorsal surface from anterior face   *Heriades*

One British species – *truncorum*. Medium-sized (7–8 mm), rather slender bee; body shining with dense, deep punctures.

–   aa   First tergite of gaster with dorsal surface smoothly rounded into anterior face   **10**

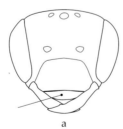

a   aa

10 (9)   a   Labrum much wider than long, easily seen even when sickle-shaped mandibles are closed   **11**

–   aa   Labrum longer than wide; in some species concealed by broad, triangular mandibles   **13**

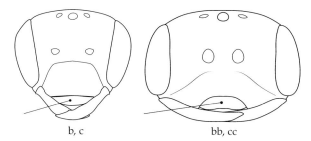

b, c   bb, cc

11 (10)   a   Marginal cell of forewing apically pointed

        b   Face narrow, eyes closer together than the height of an eye

        c   Labrum hinged below level of clypeus   **12**

   –   aa   Marginal cell of forewing apically truncate

        bb   Face broad, eyes further apart than the height of an eye

        cc   Labrum fitting into excision in lower part of clypeus so that the lower borders of each are level   ***Panurgus***

           Medium-sized species (7–9 mm); deeply black and shiny; rather hairless.

12 (11)   a   Small species, less than 7 mm

        bb   Shining black with sparse hairs

        cc   Antennal segments not longer than wide, with specialised sensory areas below

                                                                   Dufourea

           Two British species, both very rare or extinct. Resembling a small *Lasioglossum*, but with only two submarginal cells in the forewing. 5–7 mm.

   –   aa   Large species, over 12 mm

        bb   Dull black and densely hairy

        cc   Antennal segments twice as long as wide, finely keeled beneath   ***Dasypoda***

           One British species – *hirtipes*. Large (12–14 mm); banded bee with long hairs on gaster, thorax, legs and face.

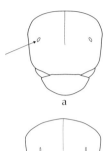

a

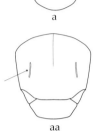

aa

13 (10) a   Parapsidal lines short, scarcely longer than wide, appearing as a raised, flattened area usually distinct from surrounding punctures (move specimen relative to light-source to create reflections) OR sternite 1 with a long ventral spine.

     b   Gaster with sternite 2 flat and unmodified

     c   Integument may be slightly metallic but black in some species   ***Osmia***

Medium-sized to large species (7–13 mm); many either with long but sparse red hair or metallic integument. 'Mason bees'.

  –   aa   Parapsidal lines linear, many times longer than wide, not always obvious

     bb   Gaster with sternite 2 modified into a raised welt

     cc   Integument always black   **14**

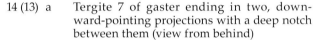

14 (13) a   Tergite 7 of gaster ending in two, downward-pointing projections with a deep notch between them (view from behind)

     b   Rim of tergite 6 with smooth outline

     c   Thorax elongate; propodeum long, with a distinct division between dorsal and posterior surfaces, the dorsal area with a series of raised longitudinal keels

               Chelostoma

Small to medium-sized species (5–11 mm); rather shining, black, elongate body.

a

aa

  –   aa   Tergite 7 of gaster ending in a single median point, this segment somewhat curved beneath the end of the gaster and partially concealed by tergite 6 (view from behind)

     bb   Rim of tergite 6 with small lateral teeth

     cc   Thorax compact; propodeum short and rounded, almost vertical with no obvious dorsal surface   ***Hoplitis***

Medium-sized species (6–10 mm); gaster shining, sparsely haired with black integument.

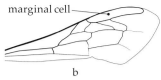

marginal cell

b

15 (1) a Surface of eyes with long dense hairs

b Marginal cell long and narrow *Apis*

The honeybee; drones infrequently found.

— aa Surface of eyes bare

bb Marginal cell usually shorter and broader **16**

16 (15) a Wings strongly purplish-iridescent

b Very large species, over 18 mm

c Body covered with dense, blackish-purple hairs *Xylocopa*

One species – *violacea* – a vagrant to Britain, but seen more often in recent years and could become established.

— aa Wings usually clear, at most smoky

bb Often smaller, if as large as 18 mm then body with bands or spots of lighter coloured hairs

cc If covered with dense, black hairs then marginal cell divided in two by an extremely fine vertical false vein and jugal lobe entirely absent **17**

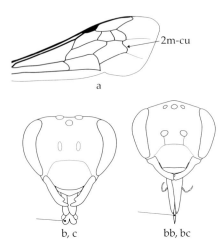

b, c   bb, bc

17 (16)   a   Vein 2m-cu strongly S-shaped, the lower end bulging outwards

         b   Tongue short and bilobed at apex

         c   Head, viewed from in front, rather triangular, the inner margins of the eyes converging ventrally   ***Colletes***

                  Medium-sized to large bees (7–15 mm), most species with dense flattened hairs covering the marginal areas of the tergites, producing a banded effect; sternum 7 modified with lateral extensions (needs to be fully visible for identification).

    –   aa   Vein 2m-cu usually straight, at most slightly curved and then not bowed outward at lower end

        bb   Tongue variable in length but always pointed at apex

        cc   Head more rounded or oval, inner margins of eyes usually more parallel   **18**

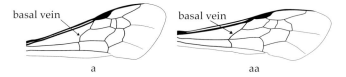

basal vein   basal vein

a   aa

18 (17)   a   Basal vein strongly arched with a distinct bend towards lower end, forming almost a right angle where it meets the longitudinal vein   **19**

    –   aa   Basal vein almost straight or slightly and evenly arched, the lower end meeting the longitudinal vein at an acute angle   **22**

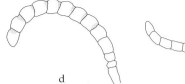 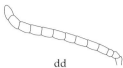

d   dd

19 (18) a   Usually, one or more tergites of gaster marked with reddish or orange-brown, shining, with sparse punctures and almost hairless - but may be completely black

         b   Face with integument black (except, perhaps, tips of mandibles)

         c   Legs black, at most tarsi translucent reddish-brown

         d   Antenna black and usually with ventral surface of apical segments with a pubescent depression basally, thus appearing 'knobbly'
              Sphecodes

              Very small to medium-sized species (4–11 mm); gaster black, usually with more-or-less extensive red belt but melanic examples occur, usually rather shining; heavily punctured head and thorax. Cleptoparasites of species of *Lasioglossum*, *Halictus* and *Andrena*.

    –   aa   Tergites usually black or dark-metallic; if red-marked then often with patches of whitish, flattened hairs and legs with yellow markings

         bb   Face nearly always with yellow markings, at least at apex of clypeus

         cc   Legs black or with yellow markings on tibia and/or tarsus

         dd   Antennae variable, sometimes yellowish-orange below, but segments usually more-or-less cylindrical and with even covering of pubescence.   **20**

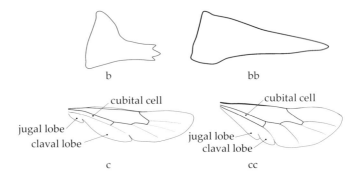

20 (19) a   Integument metallic blue AND legs black except for a minute yellow spot at base of each tibia

b   Mandibles rather short and narrowing abruptly towards apex, ending in three teeth, black

c   Jugal lobe of hindwing very short, much less than half length of claval lobe and not reaching anywhere near as far as vein closing cubital cell   ***Ceratina***

One British species – *cyanea*. Fairly small (6–7 mm), shining metallic blue bee. Nests, and overwinters, in bramble stems. The 'small carpenter bee'.

—   aa   Integument black, or if metallic then legs either entirely black or more extensively yellow

bb   Mandibles sickle-shaped, narrowing evenly from base to a simple point, sometimes yellow-marked

cc   Jugal lobe of hindwing long, well over half length of claval lobe, and extending as far as or beyond vein closing cubital cell   **21**

21 (20) a   Bands or spots of whitish flattened hairs, if present, situated on apical part of each tergite, usually extending beyond the apical margin and thus masking it; sometimes with basal bands AS WELL

b   Outer cross-veins of similar thickness and pigmentation to adjacent veins

c   IF head and thorax metallic bronze or green, then legs predominantly yellow or orange   ***Halictus***

Small to medium-sized species (6–11 mm); integument black or metallic greenish; legs extensively yellow.

	aa	Bands or spots of whitish flattened hairs, if present, situated basally and often originating beneath the apical margin of the previous tergite
	bb	Outer cross-veins thinner and less well pigmented than adjacent veins (less distinct than in females and sometimes very difficult to appreciate)
	cc	IF head and thorax metallic bronze or green, then legs dark with at most tarsi yellow

Lasioglossum

Very small to medium-sized species (4–12 mm); integument black or metallic greenish to bluish; legs usually less extensively yellow than in *Halictus*.

22 (18)	a	Face and/or mandibles with clear yellow or reddish-yellow markings on the integument **23**
–	aa	Face, and mandibles, except perhaps extreme apex, black (but may have yellow hair) **27**

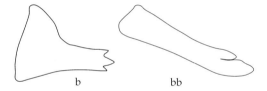

b   bb

23 (22)	a	Integument shining metallic blue
	b	Mandibles rather short and narrowing abruptly towards apex *Ceratina*

One British species – *cyanea*. Fairly small (6–7 mm), shining metallic blue bee. Nests, and overwinters, in bramble stems. The 'small carpenter bee'.

–	aa	Integument black or marked with yellow or reddish-yellow
	bb	Mandibles long, narrowing evenly from base to apex (sometimes with an accessory tooth) **24**

24 (23)	a	Some part of front or mid femora or tibiae clear yellow or reddish-yellow, usually fairly extensively (if, very occasionally, front and mid femora and tarsi apparently brownish-black then gaster with yellow spots on the integument of one or more tergites) **25**
	aa	Front and mid femora and tibiae completely black or brownish-black **26**

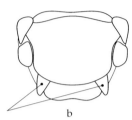

b   bb

25 (24) a Gaster with paired pale spots formed from dense flattened scale-like hairs

    b Axillae, on either side of scutellum, large and triangular, projecting backwards as a pair of teeth   ***Epeolus***

     Medium-sized species (6–11 mm); cleptoparasites of *Colletes*.

  – aa Gaster without scale-like hairs, instead patterned by yellow, red or brownish spots or bands due to pigmentation of the integument

    bb Axillae, on either side of scutellum, small and inconspicuous, not projecting backwards as teeth   ***Nomada***

     Small to large species (4–15 mm); gaster patterned and shining (hence wasp-like), head and thorax often heavily punctured. Cleptoparasites of *Andrena, Lasioglossum, Melitta* and *Eucera*.

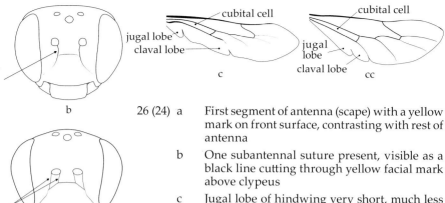

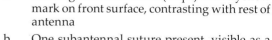

b

bb

26 (24) a   First segment of antenna (scape) with a yellow mark on front surface, contrasting with rest of antenna

             b   One subantennal suture present, visible as a black line cutting through yellow facial mark above clypeus

             c   Jugal lobe of hindwing very short, much less than half length of claval lobe (care needed – may be folded under)   **_Anthophora_**

                  Fairly to very large species (9–16 mm); usually rather hairy; gaster with hair bands in some species; eyes, in life, sometimes rather greenish.

     –   aa   First segment of antenna (scape) brownish-black, like rest of antenna

           bb   Two subantennal sutures present, showing as shining lines descending from antennal socket to meet upper border of clypeus, the shining suture defining the clypeus slightly thickened at these points

           cc   Jugal lobe of hindwing long, more than half length of claval lobe and reaching as far as the vein closing the cubital cell   **_Andrena_** (in part)

                  Fairly small to large species (6–15 mm); a small section of this large genus has the face marked with yellow; one species with tergites 2 and 3 blood-red.

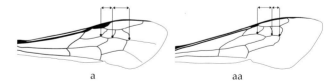

a   aa

27 (22) a   Lower border of third submarginal cell (measured between intersection with its bordering cross-veins) distinctly longer than that of second submarginal cell   **28**

     –   aa   Lower border of third submarginal cell more-or-less equal in length to that of second or even shorter   **29**

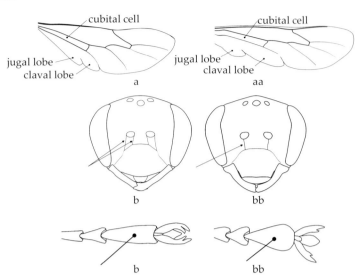

28 (27) a Jugal lobe of hindwing long; its length, measured from base of wing, distinctly more than half the length of claval lobe, usually reaching as far out as the vein closing the cubital cell (care needed – may be folded under)

b Two subantennal sutures present, showing as shining lines descending from antennal socket to meet upper border of clypeus, the shining suture defining the clypeus slightly thickened at these points

c Last tarsal segment slender, at least three times as long as wide *Andrena*

Small to large species (5–15 mm); a very large and diverse genus with species ranging from almost hairless to densely haired, shining to dull, some banded, and some with reddish markings on gaster.

— aa Jugal lobe of hindwing short, less than half length of claval lobe, and not nearly reaching vein closing the cubital cell

bb One subantennal suture present

cc Last tarsal segment broad, about twice as long as wide *Melitta*

Medium-sized to large species (8–13 mm); integument dark; some species with whitish hair bands on gaster; most species with antennal segments long and concave below, the antennae appearing 'knobbly'.

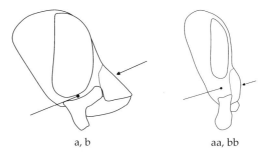

a, b   aa, bb

29 (27) a   Cheek (malar space) short, distance from lower margin of eye to mandibular base less than one fifth of width of mandible here

b   Face strongly protruding, clypeus at about 45° to vertical

c   Gaster mainly sparsely haired and shining, with very long hairs confined to base, tergites with paired lateral spots formed from flattened hairs   ***Melecta***

Large to very large bees (12–16 mm), black with greyish-white hair; scutellum with paired apical prongs (difficult to see beneath hair). Cleptoparasites of *Anthophora*.

–   aa   Cheek (malar space) long, equal to or longer than width of mandibular base

bb   Face almost flat, clypeus vertical

cc   Gaster relatively densely long-haired, especially towards apex, never with rounded spots but often with complete or interrupted bands of coloured hair   ***Bombus***

Large to very large species (12–17 mm); Bumblebees and cuckoo bumblebees. (Includes the former genus *Psithyrus* now considered a subgenus of *Bombus*).

9 References and further reading

Many journal articles are now available on the internet, either complete or as abstracts. It is worth searching there first. It is often possible to make arrangements to see or borrow books or journal articles by visiting the library of a local university, or by asking your local public library to borrow the work (or a photocopy of it) for you via the British Library Document Supply Centre (see www.bl.uk/articles). This may take some time, and it is important that your librarian, or online request, has a reference that is correct in every detail. References are acceptable in the form given here, namely the author's name, the date of publication, followed by (for a book) the title and publisher or (for a journal article) the title of the article, the journal title, the volume number, and the first and last pages of the article.

Aigner, P. A. (2006) The evolution of specialized floral phenotypes in a fine-grained pollination environment. In *Plant-Pollinator Interactions. From Specialization to Generalization*, eds. Waser, N. M. & Ollerton, J., pp. 23–46. Chicago & London: University of Chicago.

Aizer, M. A. & Harder, L. D. (2009) The global stock of domesticated honey bees is growing slower than agricultural demand for pollination. *Current Biology* **19**, 915–918.

Alcock, J. and Houston, T. (1996) Mating systems and male size in Australian Hylaeine bees (Hymenoptera: Colletidae). *Ethology* **102**, 591–610.

Andersson, M. (1984) The evolution of eusociality. *Annual Review of Ecology and Systematics* **15**, 165–190.

Ayasse, M. (1994) The meaning of male odor in the mating biology of sweat bees (Hymenoptera: Halictidae). *Les Insectes Sociaux. Proceedings of the International Congress IUSSI* **12**, 88. Paris: University of Paris Nord.

Ayasse, M. & Dutzler, G. (1998) The functions of pheromones in the mating biology of *Osmia* bees (Hymenoptera: Megachilidae). *Social Insects at the Turn of the Millennium. Proceedings of the International Congress IUSSI* **13**, 42. Adelaide: Flinders University.

Ayasse, M., Engels, W., Hefetz, A., Tengö, J. & Lübke, G. F. W. (1993) Ontogenetic patterns of volatiles identified in Dufour's gland extracts from queens and workers of the primitively eusocial halictine bee, *Lasioglossum malachurum* (Hymenoptera: Halictidae). *Insectes Sociaux* **40**, 41–58.

Ayasse, M., Engels, W., Lübke, G. & Francke, W. (1999) Mating expenditures reduced via female sex pheromone modulation in the primitively eusocial halictine bee, *Lasioglossum (Evylaeus) malachurum* (Hymenoptera: Halictidae). *Behavioral Ecology and Sociobiology* **45**, 95–106.

Ayasse, M., Paxton, R. J. and Tengö, J. (2001) Mating behaviour and chemical communication in the Order Hymenoptera. *Annual Review of Entomology* **46**, 31–78.

Baldock, D. W. (2008) *Bees of Surrey.* Woking: Surrey Wildlife Trust.

Baldock, D. W. (2010) *Wasps of Surrey.* Woking: Surrey Wildlife Trust.

Banaszak, J. (1996) Ecological basis of conservation of wild bees. In *The Conservation of Bees*, eds. Matheson, A., Buchmann, S. L., O'Toole, C., Westrich, P. & Williams, I. H., pp. 55–62. London: Academic Press.

Batra, S. W. T. & Norden, B. B. (1996) Fatty food for their broods: how *Anthophora* bees make and provision their cells (Hymenoptera: Apoidea). *Memoirs of the Entomological Society of Washington* **17**, 36–44.

Benton, T. (2006) *Bumblebees (The New Naturalist Library 98).* London: HarperCollins.

Benton, T. (2012) *Grasshoppers and Crickets (The New Naturalist Library 120).* London: HarperCollins.

Bergmark, L., Borg-Karlson, A.-K. & Tengö, J. (1984) Female characteristics and odour cues in mate recognition in *Dasypoda altercator* (Hym.: Melittidae). *Nove Acta Regiae Societas Scientiarum Uppsaliensis Series 5* **C 3**, 137–143.

Biesmeijer, J. C., Roberts, S. P. M., Reemer, M. *et al.* (2006) Parallel declines in pollinators and insect-pollinated plants in Britain and the Netherlands. *Science* **313**, 351–354.

Bogusch, P., Kratochvíl, L. & Straka, J. (2006) Generalist cuckoo bees (Hymenoptera: Apoidea: *Sphecodes*) are species-specialist at the individual level. *Behavioral Ecology and Sociobiology* **60**, 422–499.

Boomsma, J. J. (2007) Kin selection versus sexual selection: why the ends do not meet. *Current Biology* **17**, 673–683.

Borg-Karlson, A.-K., Tengö, J., Valterová, I., Unelius, C. R., Taghizade, T., Tolasch, T. and Francke, W. (2003) (S)-(+)-Linalool, a mate attractant pheromone component in the bee *Colletes cunicularius. Journal of Chemical Ecology* **29**, 1–14.

Bosch, J. & Vicens, N. (2002) Body size as an estimator of production costs in a solitary bee. *Ecological Entomology* **27**, 129–137.

Brock, P. D. (2014) *A Comprehensive Guide to Insects of Britain and Ireland.* Newbury: Pisces.

Butler, C. G. (1965) Sex attraction in *Andrena flavipes* Panzer (Hymenoptera: Apidae), with some observations on nest-site restriction. *Proceedings of the Royal Entomological Society of London. Series A: General Entomology* **40**, 77–80.

Cane, J. H. (1983) Olfactory evaluation *of Andrena* host nest suitability by kleptoparasitic *Nomada* bees (Hymenoptera: Apoidea). *Animal Behaviour* **31**, 138–144.

Cane, J. H. & Sipes, S. (2006) Characterizing floral specialization by bees. In *Plant-Pollinator Interactions. From Specialization to Generalization*, eds. Waser, N. M. & Ollerton, J., pp. 99–122. Chicago & London: University of Chicago.

Cane, J. H. & Tengö, J. (1981) Pheromonal cues direct mate-searching behaviour of male *Colletes cunicularius* (Hymenoptera: Colletidae). *Journal of Chemical Ecology* **7**, 427–436.

Carrie, R. J. G., George, D. R. & Wäckers, F. L. (2012) Selection of floral resources to optimise conservation of agriculturally-functional insect groups. *Journal of Insect Conservation* **16**, 635–640.

Chinery, M. (2005) *Complete British Insects.* London: HarperCollins.

Chittka, L. & Thomson, J. D. (eds.) (2001) *The Cognitive Ecology of Pollination: Animal behaviour and Floral Evolution.* Cambridge: Cambridge University Press.

Comba, L., Corbet, S. A., Hunt, L. & Warren, B. (1999) Flowers, nectar and insect visits: evaluating British plant species for pollinator-friendly gardens. *Annals of Botany* **83**, 369–383.

Conrad, T., Paxton, R. J., Barth, F. G., Francke, W. & Ayasse, M. (2010) Female choice in the red mason bee, *Osmia rufa* (L.) (Megachilidae). *The Journal of Experimental Biology* **213**, 4065–4073.

Corbet, S. A. (1995) Insects, plants and succession: advantages of long-term set-aside. *Agriculture, Ecosystems and Environment* **53**, 201–217.

Corbet, S. A. (1996) Which bees do plants need? In *The Conservation of Bees,* eds. Matheson, A., Buchmann, S. L., O'Toole, C., Westrich, P. & Williams, I. H., pp. 105–113. London: Academic Press.

Corbet, S. A. (2006) A typology of pollination systems: implications for crop management and the conservation of wild plants. In *Plant-Pollinator Interactions. From Specialization to Generalization,* eds. Waser, N. M. & Ollerton, J., pp. 315–340. Chicago & London: University of Chicago.

Corbet, S. A., Bee, J., Dasmahapatra, K., Gale, S., Gorringe, E. *et al.* (2001) Native or exotic? Double or single? Evaluating plants for pollinator-friendly gardens. *Annals of Botany* **87**, 219–232.

Corbet, S. A. & Huang, S. Q. (2014) Buzz pollination in eight bumblebee-pollinated *Pedicularis* species: does it involve vibration-induced triboelectric charging of pollen grains? *Annals of Botany* **114**, 1665–1674.

Corbet, S. A., Saville, N. M., Fussell, M., Prŷs-Jones, O. E. & Unwin, D. M. (1995) The competition box; a graphical aid to forecasting pollinator performance. *Journal of Applied Ecology* **32**, 707–719.

Danforth, B. N. (1990) Provisioning behavior and the estimation of investment ratios in a solitary bee, *Calliopsis (Hypomacrotera) persimilis* (Cockerell) (Hymenoptera: Andrenidae). *Behavioral Ecology and Sociobiology* **27**, 159–168.

Danforth, B. N., Cardinal, S., Praz, C., Almeida, D. and Michez, D. (2013) The impact of molecular data on our understanding of bee phylogeny and evolution. *Annual Review of Entomology* **58**, 57–78.

Davis, E. S., Murray, T. E., Fitzpatrick, U., Brown, M. J. F. & Paxton, R. J. (2010) Landscape effects on extremely fragmented populations of a rare solitary bee, *Colletes floralis. Molecular Ecology* **19**, 4922–4935

Dicks, L. V., Corbet, S. A. & Pywell, R. F. (2002) Compartmentalization in plant-insect flower visit webs. *Journal of Animal Ecology* **71**, 32–43.

Dicks, L. V., Showler, D. A. & Sutherland, W. J. (2010) *Bee Conservation: Evidence for the Effects of Conservation.* Exeter: Pelagic Publishing.

Dobson, H. E. M. (1987) Role of flower and pollen aromas in host-plant recognition by solitary bees. *Oecologia* **72**, 618–623.

Dukas, R. (2001) Effects of predation risk on pollinators and plants. In *The Cognitive Ecology of Pollination: Animal behaviour and Floral Evolution,* eds. Chittka, L. & Thomson, J. D., pp. 214–236. Cambridge: Cambridge University Press.

Dutzler, G. & Ayasse, M. (1996) The function of female sex pheromones in mate selection of *Osmia rufa* (Megachilidae) bees. *Proceedings of the 20th International Congress of Entomology*: 376.

Edwards, M. (1996) Optimizing habitats for bees in the United Kingdom - a review of recent conservation action. In *The Conservation of Bees*, eds. Matheson, A., Buchmann, S. L., O'Toole, C., Westrich, P. & Williams, I. H., pp. 35–45. London: Academic Press.

Edwards, M. (2013a) Bees in Britain – are they all important pollinators? *British Wildlife* **25**, 12–15.

Edwards, M. (2013b) *An Introduction to the Wild Bees of Scotland*. Perth: Scottish Natural Heritage.

Edwards, R. (ed.) (1997) *Provisional Atlas of the Aculeate Hymenoptera of Britain and Ireland. Part 1*. Huntingdon: Biological Records Centre/BWARS.

Edwards, R. (ed.) (1998) *Provisional Atlas of Aculeate Hymenoptera of Britain and Ireland. Part 2*. Huntingdon: Biological Records Centre/BWARS.

Edwards, R. & Broad, G. (eds.) (2005) *Provisional Atlas of the Aculeate Hymenoptera of Britain and Ireland. Part 5*. Huntingdon: Biological Records Centre/BWARS.

Edwards, R. & Broad, G. (eds.) (2006) *Provisional Atlas of the Aculeate Hymenoptera of Britain and Ireland. Part 6*. Huntingdon: Biological Records Centre/BWARS.

Edwards, R. & Roy, H. E. (eds.) (2009) *Provisional Atlas of the Aculeate Hymenoptera of Britain and Ireland. Part 7*. Wallingford: Biological Records Centre/BWARS.

Edwards, R. & Roy, H. E. (eds.) (2012) *Provisional Atlas of the Aculeate Hymenoptera of Britain and Ireland. Part 8*. Wallingford: Biological Records Centre/BWARS.

Edwards, R. & Telfer, M. G. (eds,) (2001) *Provisional Atlas of the Aculeate Hymenoptera of Britain and Ireland. Part 3*. Huntingdon: Biological Records Centre/BWARS.

Edwards, R. & Telfer, M. G. (eds.) (2002) *Provisional Atlas of the Aculeate Hymenoptera of Britain and Ireland. Part 4*. Huntingdon: Biological Records Centre/BWARS.

Ellis, J. S., Knight, M. E., Darvill, B. & Goulson, D. (2006) Extremely low effective population sizes, genetic structuring and reduced genetic diversity in a threatened bumblebee species, *Bombus sylvarum* (Hymenoptera: Apidae). *Molecular Ecology* **15**, 4375–4386.

Ellis, W. N. & Ellis-Adams, A. C. (1993) To make a meadow it takes a clover and a bee: the entomophilous flora of N. W. Europe and its insects. *Bijdragen tot de Dierkunde* **63**, 193–220.

Else, G. R. (1995a) The distribution and habits of the small carpenter bee *Ceratina cyanea* (Kirby 1802) (Hymenoptera: Apidae) in Britain. *British journal of Entomology and Natural History* **8**, 1–6.

Else, G. R. (1995b) The distribution and habits of the bee *Hylaeus pectoralis* Förster, 1871, (Hymenoptera: Apidae) in Britain. *British Journal of Entomology and Natural History* **8**, 43–47.

Else, G. R. (1998) The status of *Stelis breviuscula* (Nylander) (Hymenoptera: Apidae) in Britain, with a key to the British species of *Stelis. British Journal of Entomology and Natural History* **10**, 214–216.

Else, G. R. & Edwards, M. 2018 *Handbook of the Bees of the British Isles (vols. 1 & 2)*. London: Ray Society

Faegri, K. & van der Pijl, L. (1979) *The Principles of Pollination Ecology*. Oxford: Pergamon.

Falk, S & Lewington, R. (2015) *Field Guide to the Bees of Great Britain and Ireland*. Oxford: British Wildlife Publishing.

Field, J. (1996) Patterns of provisioning and iteroparity in a solitary halictine bee, *Lasioglossum* (*Evylaeus*) *fratellum* (Perez) with notes on *L.* (*E.*) *calceatum* (Scop.) and *L.* (*E.*) *villosulum* (K.). *Insectes Sociaux* **43**, 167–182.

Field, J., Paxton, R. J., Soro, A. & Bridge, C. (2010) Cryptic plasticity underlies a major evolutionary transition. *Current Biology* **20**, 2028–2031.

Fitzpatrick, U., Murray, T. E., Paxton, R. J. & Brown, M. J. E. (2007) Building on IUCN regional red lists to produce list of species of conservation priority: a model with Irish bees. *Conservation Biology* **21**, 1324–1332.

Fitzpatrick, U., Stout, J. *et al.* (2015) *All Ireland Pollinator Plan 2015 – 2020*. Waterford: National Biodiversity Data Centre, series 3.

Free, J. B. (1993) *Insect Pollination of Crops. 2nd edn*. London: Academic Press.

Gallai, N., Salles, J. M., Settele, J. *et al.* (2009) Economic valuation of the vulnerability of world agriculture confronted with pollinator decline. *Ecological Economics* **68**, 810–821.

Gardiner, T. (2012) How does mowing of grassland on sea wall flood defences affect insect assemblages in eastern England? In *Grassland: Types, Biodiversity and Impacts*, ed. Zhang, W.-J. USA: Nova Science.

Gardiner, T, Pilcher, R. & Wade, M. (2015) *Sea Wall Biodiversity Handbook*. Abingdon, Oxfordshire: RPS Group.

Gathmann, A. & Tscharntke, T. (2002) Foraging ranges of solitary bees. *Journal of Animal Ecology* **71**, 757–764.

Gazoul, J. (2005) Buzziness as usual? Questioning the global pollination crisis. *Trends in Ecology and Evolution* **20**, 367–373.

Gegear, R. J. & Laverty, T. M. (2001) The effect of variation among floral traits on the flower constancy of pollinators. In *The Cognitive Ecology of Pollination: Animal behaviour and Floral Evolution*, eds. Chittka, L. & Thomson, J. D., pp. 1–20. Cambridge: Cambridge University Press.

Goodell, K. (2003) Food availability affects *Osmia pumila* (Hymenoptera: Megachilidae) foraging, reproduction, and brood parasitism. *Oecologia* **134**, 318–427.

Habermannová, P., Bogusch, P. & Straka, J. (2013) Flexible host choice and common host switches in the evolution of generalist and specialist cuckoo bees (Anthophila: *Sphecodes*). *PLOS ONE* **8**(5), e64537 (open access online journal).

Hanley, M. E. & Wilkins, J. P. (2015) On the verge? Preferential use of road-facing hedgerow margins by bumblebees in agro-ecosystems. *Journal of Insect Conservation* **19**, 67–74.

Hefetz, A. (1998) Exocrine glands and their products in non-*Apis* bees: chemical, functional and evolutionary perspectives. In *Pheromone Communication in Social Insects*, eds. Vander Meer, R. K., Breed, M. D., Winston, M. L. & Espelie, K. E., pp. 236–256. Boulder: Westview.

Heinrich, B. (1979) *Bumblebee Economics*. Cambridge, Mass. & London: Harvard University Press.

Jordano, P., Bascompte, J. & Olesen, J. M. (2006) The ecological consequences of complex topology and nested structure in pollination webs. In *Plant-Pollinator Interactions. From Specialization to Generalization*, eds. Waser, N. M. & Ollerton, J., pp.

173–199. Chicago & London: University of Chicago.

Kearns, C. A., Inouye, D. W. & Waser, N. M. (1998) Endangered mutualisms: the conservation of plant-pollinator interactions. *Annual Review of Ecology and Systematics* **29**, 83–112.

Kim, J.-Y. (1999) Influence of resource level on maternal investment in a leaf-cutter bee (Hymenoptera: Megachilidae). *Behavioral Ecology* **10**, 552–556.

Kirk, W. D. J. & Howes, F. N. (2012) *Plants for Bees*. Cardiff: International Bee Research Association.

Kleijn, D., Winfree, R., Bartomeus, I. & 55 other contributors (2015) Delivery of crop pollination services is an insufficient argument for wild pollinator conservation. *Nature Communications* **6**(7414) June 16 (online journal).

Larkin, L. L., Neff, J. L. & Simpson, B. B. (2008) The evolution of a pollen diet: host choice and diet breadth of *Andrena* bees (Hymenoptera: Andrenidae). *Apidologie* **39**, 133–145.

Larsen, O. N., Gleffe, G. and Tengö, J. (1986) Vibration and sound communication in solitary bees and wasps. *Physiological Entomology* **11**, 287–296.

Matheson, A., Buchmann, S. L., O'Toole, C., Westrich, P. & Williams, I. H. (1996) *The Conservation of Bees*. London: Academic Press.

Memmott, J., Waser, N. M. & Price, M. V. (2004) Tolerance of pollination networks to species extinctions. *Proceedings of the Royal Society of London B* **271**, 2605–2611.

Menzel, R. (2001) Behavioral and neural mechanisms of learning and memory as determinants of flower constancy. In *The Cognitive Ecology of Pollination: Animal behaviour and Floral Evolution*, eds. Chittka, L. & Thomson, J. D., pp. 21–40. Cambridge: Cambridge University Press.

Michener, C. D. (2007) *The Bees of the World*. Baltimore: Johns Hopkins University Press.

Milet-Pinheiro, P., Ayasse, M., Dobson, H. E. M., Clemens, S., Francke, W. & Dötterl, S. (2013) The chemical basis of host-plant recognition in a specialised bee pollinator. *Journal of Chemical Ecology* **39**, 1347–1360.

Minckley, R. L. & Roulston, T. H. (2006) Incidental mutualisms and pollen specialization among bees. In *Plant-Pollinator Interactions. From Specialization to Generalization*, eds. Waser, N. M. & Ollerton, J., pp. 69–98. Chicago & London: University of Chicago.

Müller, A & Kuhlmann, M. (2018) Pollen hosts of western Palaearctic bees of the genus *Colletes* (Hymenoptera: Colletidae): the Asteraceae paradox. *Biological Journal of the Linnean Society* **95**, 719–733.

Nature Conservancy Council (1991) *A Review of the Scarce and Threatened Bees, Wasps and Ants of Great Britain*. Peterborough: NCC.

Neff, J. L. (2008) Components of nest provisioning behaviour in solitary bees (Hymenoptera: Apoidea). *Apidologie* **39**, 30–45.

Neff, J. L. & Simpson, B. B. (1997) Nesting and foraging behaviour of *Andrena* (*Callandrena*) *rudbeckiae* Robertson (Hymenoptera: Apoidea: Andrenidae) in Texas. *Journal of the Kansas Entomological Society* **70**, 100–113.

Ollerton, J., Winfree, R. & Tarrant, S. (2011) How many flowering plants are pollinated by animals? *Oikos* **120**, 321–326.

O'Toole, C. (1993) Diversity of native bees and agroecosystems.

In *Hymenoptera and Biodiversity*, eds. LaSalle, J. & Gauld, I. D.. Wallingford, UK: CAB International.

O'Toole, C. (2012) Plants for solitary bees. In *Plants for Bees*, eds. Kirk, W. D. J. & Howes, F. N., chapter 4. Cardiff: International Bee Research Association.

O'Toole, C. (2013) *Bees: A Natural History*. New York & Ontario: Firefly.

Parker, R. (2008) Observations of a mason bee, *Osmia bicolor*, concealing its nest in a snail shell (7–24th May, 2008). *Suffolk Natural History* **44**, 51–52.

Paxton, R. J. (2005) Male mating behaviour and mating systems in bees: an overview. *Apidologie* **36**, 145–156.

Paxton, R. J., Ayasse, M., Field, J. & Soro, A. (2002) Complex sociogenetic organization and reproductive skew in a primitively eusocial sweat bee, *Lasioglossum malachurum*, as revealed by microsatellites. *Molecular Ecology* **11**, 2405–2416.

Paxton, R. J., Kukuk, P. F. & Tengo, J. (1999) Effects of familiarity and nestmate number on social interactions in two communal bees, *Andrena scotica* and *Panurgus calcaratus* (Hymenoptera, Andrenidae). *Insectes Sociaux* **46**, 109–118.

Paxton, R. J & Pohl, H. (1999) The tawny mining bee, *Andrena fulva* (Müller) (Hymenoptera, Andreninae) at a South Wales field site and its associated organisms: Hymenoptera, Diptera, Nematoda and Strepsiptera. *British Journal of Entomology and Natural History* **12**, 57–67.

Paxton, R. J. & Tengö, J. (1996) Internidal mating, emergence and sex ratio in a communal bee *Andrena jacobi* Perkins 1921 (Hymenoptera: Andrenidae). *Journal of Insect Behavior* **9**, 421–440.

Paxton, R. J., Thorén, P. A., Tengö, J., Estoup, A. & Pamilo, P. (1996) Mating structure and nestmate relatedness in a communal bee, *Andrena jacobi* (Hymenoptera: Andrenidae), using microsatellites. *Molecular Ecology* **5**, 511–519.

Perkins, V. R. (1884) On a singular habit of *Osmia bicolor* Sch. *Entomologists' Monthly Magazine* **21**, 67–68.

Potts, S. G., Biesmeijer, J. C., Kremen, C., Neumann, P., Schweiger, O. & Kunin, W. E. (2010) Global pollinator declines: trends, impacts and drivers. *Trends in Ecology and Evolution* **25**, 345–353.

Proctor, M., Yeo, P. & Lack, A. (1996) *The Natural History of Pollination (The New Naturalist Library 83)*. London: HarperCollins.

Raguso, R. A. (2001) Floral scent, olfaction and scent-driven foraging behaviour. In *The Cognitive Ecology of Pollination: Animal behaviour and Floral Evolution*, eds. Chittka, L. & Thomson, J. D., pp. 83–105. Cambridge: Cambridge University Press.

Raw, A. (1972) The biology of the solitary bee *Osmia rufa* (L.) (Megachilidae) *Transactions of the Royal Entomological Society of London* **24**, 213–229.

Richards M. H., French, D. & Paxton, R. J. (2005) It's good to be queen: classically eusocial colony structure and low worker fitness in an obligately social sweat bee. *Molecular Ecology* **14**, 4123–4133.

Ricketts, T. H., Regetz, J., Steffan-Dewenter, I. *et al.* (2008) Landscape effects on crop pollination services: are there general patterns? *Ecology Letters* **11**, 499–515.

Saunders, E. (1896) *The Hymenoptera Aculeata of the British Islands*. London: L. Reeve & Co.

Schiestl, F. P. & Ayasse, M. (2000) Post-mating odor in females of the solitary bee, *Andrena nigroaenea* (Apoidea, Andrenidae), inhibits male mating behaviour. *Behavioural Ecology and Sociobiology* **48**, 303–307.

Schiestl, F. P., Ayasse, M., Paulus, H. F., Löfstedt, C., Hansson, B. S., Ibarra, F., Francke, W. (2000) Sex pheromone mimicry in the early spider orchid (*Ophrys sphegodes*): patterns of hydrocarbons as the key mechanism for pollination by sexual deception. *Journal of Comparative Physiology A* **186**, 567–574.

Sedivy, C., Müller, A. & Dorn, S. (2011) Closely related pollen generalist bees differ in their ability to develop on the same pollen diet: evidence for physiological adaptations to digest pollen. *Functional Ecology* **25**, 718–725.

Seidelmann, K. (2006) Open-cell parasitism shapes maternal investment patterns in the Red Mason Bee (*Osmia rufa*). *Behavioral Ecology* **17**, 839–848.

Seidelmann, K., Ulbrich, K., and Mielenz, N. (2010) Conditional sex allocation in the Red Mason bee, *Osmia rufa*. *Behavioral Ecology and Sociobiology* **64**, 337–347.

Severinghaus, L. L., Kurtak, B. H. and Eickwort, G. C. (1981) The reproductive behaviour of *Anthidium manicatum* (Hymenoptera: Megachilidae) and the significance of size for territorial males. *Behavioral Ecology and Sociobiology* **9**, 51–58.

Sick, M., Ayasse, M., Tengö, J., Engels, W., Lübke, G. & Franke, W. (1994) Host-parasite relationships in six species of *Sphecodes* bees and their halictid hosts: nest invasion, intranidal behavior, and Dufour's gland volatiles (Hymenoptera: Halictidae). *Journal of Insect Behavior* **7**, 101–117.

Smith, B. H. & Weller, C. (1989) Social competition among gynes in halictine bees: the influence of bee size and pheromones on behavior. *Journal of Insect Behavior* **3**, 397–411.

Smith, R. M., Gaston, K. J., Warren, P. H. & Thompson, K. (2006b) Urban domestic gardens (VII): environmental correlates of invertebrate abundance. *Biodiversity and Conservation* **15**, 2515–2545.

Smith, R. M., Warren, P. H., Thompson, K. & Gaston, K. J. (2006a) Urban domestic gardens (VI): environmental correlates of invertebrate species richness. *Biodiversity and Conservation* **15**, 2415–2438.

Soro, A., Ayasse, M., Zobel, M. U. & Paxton, R. J. (2009) Complex sociogenetic organization and the origin of unrelated workers in a eusocial sweat bee, *Lasioglossum malachurum*. *Insectes Sociaux* **56**, 55–63.

Stebbins, G. L. 1970 Adaptive radiation of reproductive characteristics in angiosperms. 1. Pollination mechanisms. *Annual Review of Ecology and Systematics* **1**, 307–326.

Steffan-Dewenter, I., Klein, A.-M., Gaebele, V., Alfert, T. & Tscharntke, T. (2006) Bee diversity and plant-pollinator interactions in fragmented landscapes. In *Plant-Pollinator Interactions. From Specialization to Generalization*, eds. Waser, N. M. & Ollerton, J., pp. 387–487. Chicago & London: University of Chicago.

Steffan-Dewenter, I., Münzenberg, U., Bürger, C., Thies, C. & Tscharntke, T. (2002) Scale-dependent effects of landscape structure on three pollinator guilds. *Ecology* **83**, 1421–1432.

Steffan-Dewenter, I., Potts, S. G. & Packer, L. (2005) Pollinator diversity and crop pollination services are at risk. *Trends in Ecology and Evolution* **20**, 651–652.

Stone, G. N. (1993) Endothermy in the solitary bee *Anthophora plumipes* (Hymenoptera: Anthophoridae): independent measures of thermoregulatory ability, costs of warm-up and the role of body size. *Journal of Experimental Biology* **174**, 299–320.

Stone, G. N. (1995) Female foraging responses to sexual harassment in the solitary bee *Anthophora plumipes*. *Animal Behaviour* **50**, 405–412.

Stone, G. N., Loder, P. M. J., and Blackburn, T. M. (1995) Foraging and courtship behaviour in males of the solitary bee *Anthophora plumipes* (Hymenoptera: Anthophoridae): thermal physiology and the role of body size. *Ecological Entomology* **20**, 169–183.

Stone, G. N. and Willmer, P. G. (1989) Warm-up rates and body temperatures in bees: the importance of body size, thermal regime and phylogeny. *Journal of Experimental Biology* **147**, 303–328.

Strudwick, T. (2011) The bees of Norfolk: a provisional county list. *Transactions of the Norfolk and Norwich Naturalists' Society* **44**, 34–57.

Stubbs, A. & Drake, M. (2014) *British Soldierflies and their Allies. 2nd edition.* Reading: British Entomological and Natural History Society.

Tengö, J. (1979) Odour-releasing behaviour in *Andrena* male bees. *Zoon* **7**, 15–48.

Tengö, J., Ågren, L., Bauer, B., Isaksson, R., Liljefors, T. *et al.* (1990) *Andrena wilkella* male bees discriminate between enantiomers of cephalic secretion components. *Journal of Chemical Ecology* **16**, 429–441.

Tengö, J. & Bergström, G. (1976) Comparative analysis of lemon-smelling secretions from heads of *Andrena* F. (Hymenoptera: Apoidea) bees. *Comparative Biochemistry and Physiology B* **57**, 179–188.

Tengö, J. & Bergström, G. (1977a) Comparative analyses of complex secretions from heads of *Andrena* bees (Hym., Apoidea) *Comparative Biochemistry and Physiology B* **57**, 197–202.

Tengö, J. & Bergström, G. (1977b) Cleptoparasitism and odor mimetism in bees: do *Nomada* males imitate the odor of *Andrena* females? *Science* **196**, 1117–1119.

Tengö, J., Erikkson, J., Borg-Karlson, A.-K., Smith, B. H. and Dobson, H. (1989) Mate-locating strategies and multi-modal communication in male mating behaviour of *Panurgus banksianus* and *P. calcaratus* (Apoidea: Andrenidae). *Journal of the Kansas Entomological Society* **61**, 338–395.

Vanbergen, A. J. *et al.* (Insect Pollinators Initiative) (2013) Threats to an ecosystem service: pressures on pollinators. *Frontiers in Ecology and the Environment* **11**, 251–259.

Waser, N. M. & Ollerton, J. (eds.) (2006) *Plant-Pollinator Interactions. From Specialization to Generalization.* Chicago & London: University of Chicago.

Westphal, C., Steffan-Dewenter, I & Tscharntke, T. (2003) Mass flowering crops enhance pollinator densities at a landscape scale. *Ecology Letters* **6**, 961–965.

Westrich, P. (1996) Habitat requirements of central European bees and the problems of partial habitats. In *The Conservation of Bees*, eds. Matheson, A., Buchmann, S. L., O'Toole, C., Westrich, P. & Williams, I. H., pp. 1–16. London: Academic Press.

Westrich, P. & Schmidt, K. (1987) Pollen analysis, an auxiliary tool to study the collecting behaviour of solitary bees. *Apidologie* **18**, 199–213.

Williams, I. H. (1996) Aspects of bee diversity and crop pollination in the European Union. In *The Conservation of Bees*, eds. Matheson, A., Buchmann, S. L., O'Toole, C., Westrich, P. & Williams, I. H., pp. 63–80. London: Academic Press.

Williams, P. H. (1982) The distribution and decline of British bumblebees (*Bombus* Latr.) *Journal of Apicultural Research* **21**, 236–245.

Wittmann, D. and Blochtein, B. (1995) Why males of leafcutter bees hold the females' antennae with their front legs while mating. *Apidologie* **26**(3), 181–196.

Index

Page numbers in *italics* denote figures and in **bold** denote glossary items and definitions of terms.